The Open University

Science: a third level course

LIVING PROCESSES

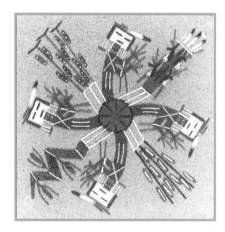

Book 4
Processes of Heredity

Edited by Steve Jones and Kay Taylor

The S327 Course Team

Course Team Chair
Steven Rose

General Editor
Mae-Wan Ho

Academic Editors
Anna Furth (Book 1)
Mae-Wan Ho (Book 2)
Steven Rose (Book 3)
Steve Jones (University College London)
(Book 4)
Kay Taylor (University College London)
(Book 4)

Authors
Mary Archer (Imperial College) (Book 2)
Jim Barber (Imperial College) (Book 2)
Samantha Bevan (Book 3)
Brian Challis (Book 1)
Basiro Davey (Book 3)
Donald Edmonds (Oxford University)
(Book 3)
Anna Furth (Book 1)
Catherine Gale (University College
London) (Book 4)
Brian Goodwin (Book 3)
David Harris (Oxford University) (Book 2)
Roger Hill (Books 1 and 2)
Mae-Wan Ho (Book 2)
Tim Hubbard (MRC Cambridge) (Book 1)
Jim Iley (Book 2)
Steve Jones (University College London)
(Book 4)
Judith Metcalfe (Book 4)
Steven Rose (Books 1 and 3)
Kay Taylor (University College London)
(Book 4)
Jonathan Wolfe (University College
London) (Book 4)

Course Manager
Jennie Simmons

Editors
Sheila Dunleavy
Gillian Riley
Bina Sharma
Margaret Swithenby

Design Group
Diane Mole (Designer)
Pam Owen (Graphic Artist)
Naz Vohra (Pagination)

BBC
Rissa de la Paz
Phil Gauron

External Course Assessor
Professor R. J. P. Williams (Oxford
University)

First published 1995

Edited, designed and typeset by The Open University.

Printed in the United Kingdom by Henry Ling Limited, The Dorset Press, Dorchester, Dorset DT1 1HD.

ISBN 0 7492 51565

This text forms part of an Open University Third Level Course. If you would like a copy of *Studying with The Open University*, please write to the Central Enquiry Service, PO Box 200, The Open University, Walton Hall, Milton Keynes, MK7 6YZ. If you have not enrolled on the Course and would like to buy this or other Open University material, please write to Open University Educational Enterprises Ltd, 12 Cofferidge Close, Stony Stratford, Milton Keynes, MK11 1BY, United Kingdom.

1.1

Contents

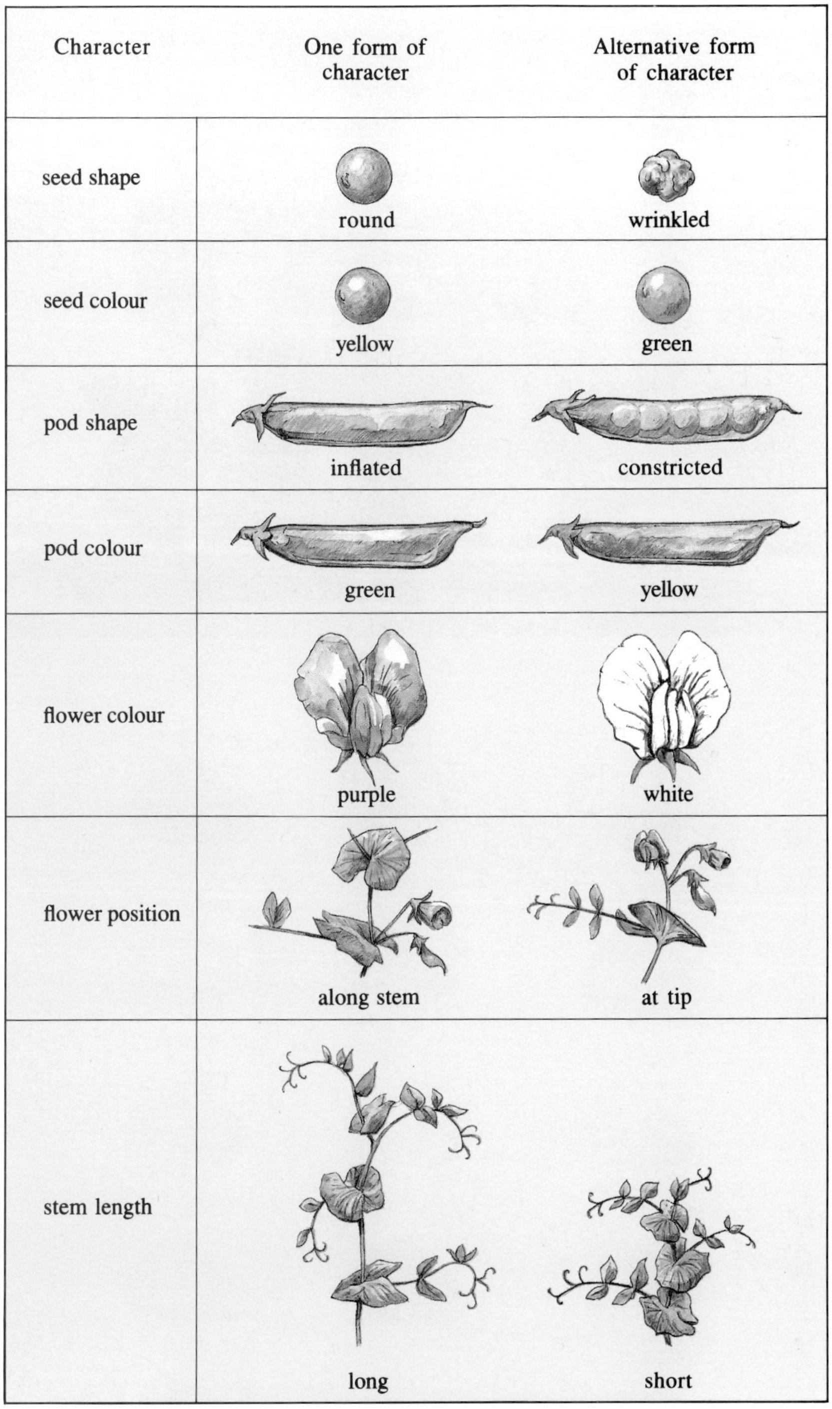

Character	One form of character	Alternative form of character
seed shape	round	wrinkled
seed colour	yellow	green
pod shape	inflated	constricted
pod colour	green	yellow
flower colour	purple	white
flower position	along stem	at tip
stem length	long	short

Plate 1.1 The seven character differences in peas studied by Mendel.

Plate 2.1 Fluorescence *in situ* hybridization. In this photograph the chromosomes are stained red; the probe (labelled with biotin) is detected with a chemical that fluoresces yellow, and can be seen hybridizing to the short arm of both copies of one of the chromosomes (this is, in fact, chromosome 3).

Plate 3.1 The tortoiseshell cat. The characteristic patterns of fur colour are the result of X inactivation.

Plate 5.1 Polymorphic land snails, *Cepaea nemoralis*.

The machinery of inheritance

1.1   Introduction

Genes are at the centre of biology. From animal behaviour to the origin of cancer, biologists use genetics in their research. The public has large – and perhaps overstated – expectations of what genetics will soon be able to do. In this chapter we review its history; the misguided origins of human genetics (which was initially seen as a branch of sociology) and the parallel development of an authentic science of heredity which began with Mendel. We remind ourselves of his work, of the chromosome theory and of how Mendel's laws apply to human inheritance. The complexities of modern genetics are illustrated by a molecular microcosm: the human globin genes. Errors in these are the most frequent cause of inborn disease in the world and the mass of information on haemoglobin shows just how complicated even an apparently simple inherited attribute can be when studied in detail.

1.2   Genes, expectations and realities

Why study genetics? Understanding inheritance is, of course, central to biology. There are also more practical reasons for asking how information is transmitted between generations. In the developed world, improved sanitation and the control of infectious disease mean that the biggest killers of children lie in the genes; and, at least in relatively affluent countries such as Britain, two people out of every three die for reasons connected, directly or indirectly, with the genes they carry.

In addition, there is the promise of fortune concealed in the DNA. Genetic engineering may make it possible to create new crops; for example, tomatoes containing frost-resistance genes from fish, or sheep producing human growth hormone in their milk. Agricultural products of genetic engineering predicted for the next decade include corn that makes its own fertilizer and pigs twice the size of those today. New drugs targeted only towards diseased cells (such as those in the early stages of cancer) may emerge, and there should soon be engineered vaccines which, by incorporating genes from several pathogens, will be able to protect against many infections at once. Perhaps, in time, the whole of medicine will be tailored to the individual needs of a patient, as determined by the genes they carry.

For these and other reasons genetics is now more prominent in the public mind than at any time in its brief history. Its potential is encapsulated in the **Human Genome Project**: the scheme to sequence the 3 000 million DNA bases which provide the biological instruction manual to make a human being. There are only about 10 g of DNA in a single man or woman, but the complete genetic alphabet will – to use one of many tired analogies – fill 1 000 substantial telephone directories (or a single compact disc). All being well, this project should be complete early in the 21st century.

Like any new science, molecular genetics is in danger of having its potential exaggerated. It is useful to begin this book with a word of warning; most diseases, now and for the foreseeable future, will be treated in conventional ways, nearly all food will be produced by traditional means, and the majority of drugs will be manufactured in ordinary chemical plants. Most of all, medicine will continue to be – as it has been for the the greater part of its history – an empirical craft whose roots lie in practical experience rather than in pure science. It is worth remembering that most premature morbidity and mortality is *not* caused by genetic dysfunction, but rather by the unequal distribution of resources coupled with the hazards of war and social conflict.

Most geneticists are well aware of what their science cannot do. Gene therapy of inborn disease is still in its earliest stages, and the idea that biologists can improve the quality of unborn children is a fantasy. More important, it is very unlikely that – whatever secrets lie hidden in the DNA – genetics will do much to alter our perceptions of what it means to be human. Because it takes so much more than DNA to make a human being, fundamental disagreements about nature, nurture and the human condition will continue long after the Genome Project is complete. In 1859, the year of publication of Charles Darwin's *The Origin of Species*, there were dire warnings of calamity because our vision of ourselves had been destroyed. Most Victorians were, in fact, remarkably indifferent to the fact that humans are part of the animal world. In the same way, most people today are likely to remain unmoved by whatever emerges from the map of the genes.

However, a few geneticists and many more members of the public continue to believe that genetics can accomplish more than it is in fact able to. Some are optimists; they believe that inherited diseases can be treated by replacing damaged DNA, or even that the prospect of 'designer babies' – whose intelligence, appearance and personality is planned by their parents – is on the horizon. For others, genetics and fear go together. There is something about the study of inheritance that touches on our deepest anxieties. One hundred and fifty years after Mary Shelley created Dr Frankenstein, his monster is being re-invented as a hackneyed but potent image with which to berate geneticists.

Genetics is, more than most, a science with a past. Although it may be overstated, public concern about what biology may soon be able to do is not, in general, irrational. There are understandable worries about our individual destiny or that of our children, together with apprehensions about the dangers of new organisms created by genetic engineering. Most people, though, are happy to leave the long-term prospects of the human race to fate – or to physics and chemistry, which have produced so many ingenious means of destruction that nowadays there seems almost no point in worrying about the distant biological future.

Such nonchalance about unborn generations is new. The primary concern of human genetics was, for many years after it began, with just these issues: about what must be done to preserve the evolutionary future and to save humanity from decline. The fate – and rights – of individuals often seemed unimportant when compared with such grand concerns.

Geneticists are well advised to show a certain scepticism about the wilder claims made on behalf of their own science. Their mistrust has a foundation in history. A century ago, human genetics had its own somewhat misconceived beginning. It can be traced to Francis Galton, Darwin's cousin. Galton argued in his 1869 book *Hereditary Genius* that people of high 'biological quality' – geniuses (amongst whom, of course, he counted himself) – were having fewer children than those genetically less well endowed. With a typically Victorian urge to interfere, he advocated a policy to prevent this decay of the human race. The science (or pseudo-science) of **eugenics** had as its goal 'to check the birth rate of the unfit and improve the race by furthering the productivity of the fit by early marriages of the best stock'.

The history of the eugenics movement is too well known to need recounting here. It led directly to the Nazi sterilization programme and to its equivalents which saw thousands of Americans sterilized in the 1930s. United States immigration laws were based for more than half a century on a eugenic attempt to keep those alleged to have poor heredity out of the country. Even today there are arguments, from both ends of the political spectrum, that it is our duty to improve our genetic heritage; and, unpalatable though it seems, decisions about genetic quality are made whenever parents at risk of having a child with inborn disease go for genetic counselling. History continues to pursue the study of human inheritance. In Germany, for example, there is still resistance to genetic screening and even to the use of genetically manipulated crops fueled by memories of what was done in the name of science a mere half-century ago.

All this means that human genetics is a science surrounded more than most with moral questions. It is not the function of this book to pursue them but it is foolish to deny that they exist. Many of these issues are touched upon in *Case Study 4* which accompanies this book.

1.3   A brief history of inheritance: 10 000 years in 700 words

People have long been aware of heredity, the tendency for offspring to resemble their parents. Much of the Old Testament is, after all, nothing more than a series of pedigrees. There are references to selective breeding of crops in Assyrian inscriptions; and there has been some understanding of inheritance since the beginnings of farming 10 000 years ago. However, an interest in the genetic material itself is a novel thing.

For most of history it was widely assumed that inheritance involved **blending**. It used to be thought that characters from each parent pass to their blood, which is mixed in the offspring. Some patterns of inheritance seem to support this idea. For example, a tall mother and a short father often produce a child of intermediate height.

Blending, though, fails to explain other aspects of heredity. Sometimes, a child looks quite different from its parents – nearly all albino individuals, for example, are the offspring of perfectly normal people and the majority of children with inherited disease are born into families who have no record of genetic illness. Sometimes a child looks more like an ancestor, or a distant relative. More damning for the idea of blending is that, if genetics were indeed based on the mixing of fluids, differences between individuals would decrease over successive generations: everyone would, in time, be average and there would be no variation at all.

Galton tested the theory by transfusing blood from a black rabbit to a white one, with the idea that if characters passed to the blood of each parent and then mixed, the recipient should have black (or perhaps grey) offspring. It did not.

There was also an acceptance (which can be traced to the great French biologist Lamarck, among others) that characters acquired during a lifetime would affect the nature of the offspring. This, too, has long been dismissed by orthodox biology (although, perhaps predictably, some of the new molecular discoveries to be described later in this book fit quite well into a Lamarckian model of inheritance – and it is worth remembering that although Lamarck is usually derided in the world of English-speaking science, his ideas on inheritance were not very different from those of Darwin).

The 19th century German, August Weissmann, drew the crucial distinction which put paid to the Lamarckian hypothesis. He suggested that, starting from the fertilized egg, there are two processes of cell division. One leads to the body (the **soma**), the other to the **germ line** – the **gametes** which give rise to the next generation. The soma will die, but the germ line is potentially immortal. If the two are independent, then characters acquired by the soma cannot be passed through the germ line.

Weissman was the first to realize that it is the passage of *information*, rather than of material, between generations that is crucial. We now know that somatic cells follow genetic rules of their own (as you will see in Chapter 7), but Weissman's doctrine is still at the centre of modern genetics.

Francis Crick, the co-discoverer of the structure of DNA, rephrased it in molecular terms in the 1960s. He argued that a one-way flow of information, from DNA through RNA to proteins, was fundamental to the relationship between genes and characters. The information is encoded in the base sequence of DNA and is copied and transmitted to subsequent generations through the process of DNA replication. It is extensive enough to code for the means of its own reproduction – which involves, among other things, constructing a living body, the soma. Crick called this the **'central dogma'** of molecular biology (although, later in his career, Crick made the engaging admission that he had not actually known what the word 'dogma' meant when he came up with the phrase) and it became a persuasive reassertion of a 19th century doctrine in 20th century terms.

Unknown to Darwin, Weissman or Galton, the separation between the inherited message and its product, the organism, had already been established. Gregor Mendel's paper of 1865 on patterns of inheritance in peas was overlooked for 35 years after its publication. Mendel's work marks the beginning of genetics as a science. As Mendelism remains central to the whole of genetics it is worth reminding ourselves of what Mendel did.

1.4   Mendel and the beginnings of genetics

Mendel's breakthrough was simple: to count rather than to measure. He was the first to see the importance of studying discrete characters whose numbers could be tallied up each generation. Mendel used peas, for several reasons. Most important, they existed as **pure lines** which had been kept separate for many generations. Each line differs from others in several characters – pea shape or colour, plant height, and flower colour, for example. Within a line all individuals are identical. The offspring of, say, two plants with round yellow peas from within the same pure line all have round peas, yellow in colour. Mendel studied seven different characters altogether; first one at a time, and then in groups (Plate 1.1).

His approach was to count the numbers of offspring of each type in crosses made between lines. He took advantage of a sexual peculiarity of peas and of many other plants. They are **hermaphrodites**, bearing both male and female organs on the same plant, and are **self-compatible**; egg cells on a single plant can be fertilized by pollen from the same individual.

Mendel looked first at contrasting single characteristics – for example, yellow versus green colour of the peas (i.e. seeds) themselves. He found that when, in the **parental generation** (P_1), pollen from a plant with yellow peas was used to fertilize the eggs from a plant whose peas were green, all the offspring (the **first filial generation, F_1**) had yellow peas.

This in itself was an important result. Offspring resembled one of their parents but not the other. Heredity did not, after all, depend on mingling the characteristics of parents so that offspring were the average of the two.

Mendel's next step was to cross the yellow F_1 offspring among themselves. Eggs could be fertilized with pollen from the same plant. Again, there was a surprise. In the succeeding generation (the **second filial generation, F_2**) plants with green peas reappeared. Although its effects had been masked, whatever had made the peas look green surfaced unchanged after one generation of oblivion. The instructions had been preserved, although their vehicle – the yellow-seeded F_1 plant – had altered in appearance.

Counting the peas said more. In this second generation, there was always a ratio of three yellow peas to one green.

From this simple pattern, Mendel formulated what became the basis of the whole of genetics: the separation of the outward form of the organism – its **phenotype**, from some internal and (to him) invisible elements which are the transmitters of inheritance – collectively referred to as the **genotype**. Inheritance was, he showed, based on discrete particles, passed unchanged down the generations. As we shall see (Section 1.4.2), soon after his work was rediscovered the term *gene* was coined to describe his units of inheritance.

Mendel suggested that pollen and egg carried one copy each of the instruction to produce a pea of a particular colour. A fertilized egg, or **zygote**, had two copies; one from each parent. The inherited particles, or **alleles**, were of two different types. For some, **dominants**, only one copy was needed to produce an individual with the appropriate phenotype. For others, **recessives**, two copies of the same particle were required. Plants could have two copies of the same allele (they were **homozygotes**), or one copy of each type (in which case they were **heterozygotes**). A heterozygote with one dominant and one recessive allele looked just like a dominant homozygote, although its genes were different. This meant, as Mendel showed, that the yellow plants in the F_2 were of two types: homozygotes (themselves pure-breeding) and heterozygotes (producing both yellow and green offspring). The 3 : 1 phenotypic ratio in this generation actually represents a genetic ratio of one dominant homozygote : two heterozygote : one recessive homozygote.

Homozygotes produced only one type of gamete (pollen or egg), while heterozygotes produced two, each bearing one of the alternative alleles. The two alleles for pea colour were present at an unspecified site in the plant, which Mendel called the pea colour **locus**. His interpretation is shown in Figure 1.1.

Mendel repeated his experiments on lines differing in other characters. Each cross gave exactly the same kind of result as that obtained with pea colour. For pea shape, round was dominant to wrinkled; for flower colour, purple was dominant to white; and for plant height, tall was dominant to short. There was a pattern to heredity, which had nothing to do with the actual character being inherited.

From these patterns of transmission Mendel formulated his *first law*, the law of **segregation**: characters are controlled by pairs of alleles, which segregate from each other during the formation of gametes and are restored at fertilization. The crucial point is that of segregation: dominant and recessive alleles can be present in the same plant, but *do not blend*. They separate unchanged when the male or female gametes are formed.

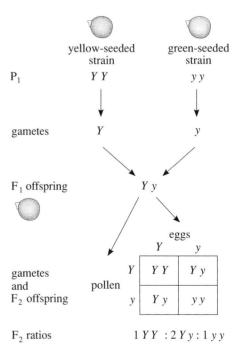

Figure 1.1   Mendel's interpretation of his first cross.

1.4.1   Inheritance at more than one locus

Mendel's pure lines differed from each other in more than one character at a time. One line might have, for example, yellow round peas on a tall plant with purple flowers, while another was characterized by green and wrinkled peas, reduced height and white flowers. He went on to consider inheritance when such lines were crossed together and very soon found that the patterns of transmission for different characters – height and colour, say – seemed to be independent.

▷   Consider a plant from a pure line with round and yellow peas crossed to one from a line with wrinkled green peas. What will be the phenotype and the genotype of the F$_1$ plants if the characters are inherited independently?

▶   There are two loci (pea shape and pea colour) involved here: the plants will be heterozygous at each of them and will hence all have the phenotype associated with the dominant alleles – that is, round and yellow seeds.

Mendel then went on to study the inheritance of many characters considered in pairs in this way; in other words, he performed **dihybrid** crosses. Figure 1.2

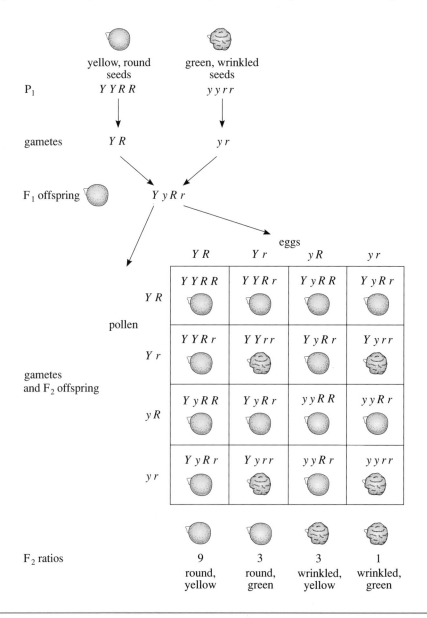

Figure 1.2   Mendel's interpretation of a dihybrid cross.

shows the results of the cross between a plant with yellow and round peas and one with green and wrinkled peas.

Mendel did the obvious experiment: self-fertilizing ('selfing') these F_1 plants, that is, combining pollen and egg from the same plant.

The F_2 of the dihybrid cross contained four different kinds of plant. Two were like those in previous generations: they had round and yellow peas or wrinkled and green. However, the other two kinds showed new arrangements of phenotypes: round with green and wrinkled with yellow. Two novel **recombinants** had been produced.

Mendel suggested that this reflected independent patterns of inheritance of alleles at the loci for shape and for colour.

From his law of segregation, we expect the F_2 ratios for each trait – shape and colour – separately to be 3 : 1; that is, 3/4 round to 1/4 wrinkled and 3/4 yellow to 1/4 green. If he was right, this provides an expectation for ratios of the two characters combined against which the results of experimental crosses of the kind carried out by Mendel can be tested.

▷   If the traits are indeed separately inherited, what – given the laws of probability – do we expect the ratios of round yellow, round green, wrinkled yellow, and wrinkled green in the F_2 to be? (*Hint*: What is the chance of getting two 'heads' if two pennies are spun at the same time?)

►   For the pennies, given that the chance of getting heads for each one independently is 1/2, and that this chance for the second penny is not altered by which way up the first one lands, the chance of two heads simultaneously is $1/2 \times 1/2$, or 1/4. In the same way, for the two separate gene loci, the expected proportions for each can be multiplied. This gives 9/16 ($3/4 \times 3/4$) round yellow plus 3/16 ($3/4 \times 1/4$) round green plus 3/16 ($3/4 \times 1/4$) wrinkled yellow plus 1/16 ($1/4 \times 1/4$) wrinkled green.

Mendel found – summed over many experiments – just this ratio in the F_2. There were (on average) nine round yellow to three round green to three wrinkled yellow to one wrinkled green (Figure 1.2). This ratio, slightly complex though it appears, can be explained in the same simple terms as those used to deduce the basic rules of inheritance. As we have just seen, the 9 : 3 : 3 : 1 ratio in the F_2 represents two 3 : 1 ratios combined at random; and, in his crosses, whatever pairs of characters he used, Mendel obtained a ratio close to this. He inferred that two gene pairs in a dihybrid cross must indeed behave independently.

From this, Mendel formulated his *second law*, that of **independent assortment**; that during the formation of gametes patterns of segregation of alleles at one locus are independent of those at other loci.

Inheritance was, it seemed, remarkably simple; depending only on the reshuffling of alternative forms of the same gene as one generation succeeds another.

Mendel was, fundamentally, right. His laws explain the patterns of inheritance of thousands of characteristics in hundreds of organisms. The fact that we now know many exceptions to his simple rules does not diminish their importance.

Mendel's laws became the touchstone against which breeding experiments could be tested. Exceptions often turned out to provide new insights into what genes were and how they worked. The interplay between theory and experiment in genetics launched a revolution in biology. Only in the past few years has genetics gone beyond Mendelism to an era in which the machinery of inheritance can be examined directly, rather than by inferring its workings from the products it makes.

1.4.2 Genes and chromosomes

The term 'gene' was coined by Wilhelm Johannsen in the early years of this century to describe Mendel's particles of inheritance. He saw them as an abstract concept rather than a physical entity. All breeding experiments depend, of course, on phenotypic characters, and Johannsen cautioned against any suggestion that there must be a simple relationship between gene and character. This has proved to be quite justified. A gene may have more than one effect on a phenotype; a phenotype may be affected by more than one gene; and genes do not work in isolation from each other or – crucially – from the environment in which they are placed.

Where might Johannsen's intangible 'genes' actually be? Some obvious clues lie within the structure of the cell. When cells divide to form gametes, the chromosomes behave in a way remarkably similar to that inferred by Mendel for his particles: their number is halved in sperm and egg compared to body cells (Figure 1.3). What is more, each chromosome behaves independently of the others when it passes to opposite poles of the dividing cell. Perhaps genes are on chromosomes. This theory was confirmed by an ingenious breeding experiment which has, as we shall see, its echoes in modern molecular genetics.

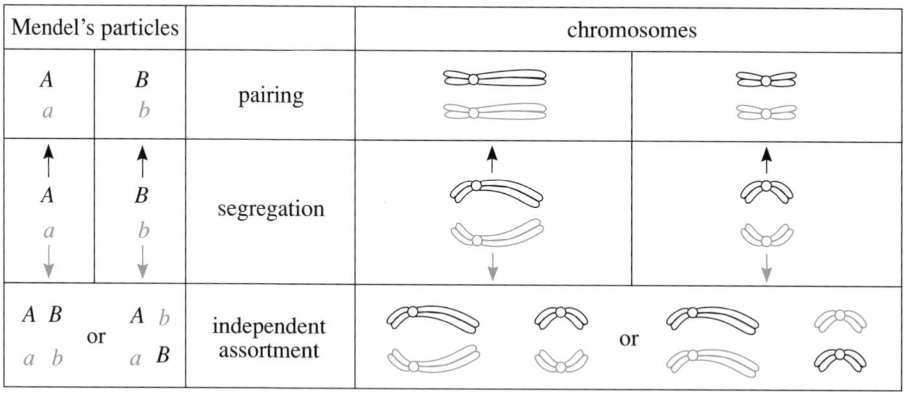

Mendel's particles			chromosomes	
A B a b		pairing		
A B a b		segregation		
A B or A b a b a B		independent assortment		or

Figure 1.3 Parallels in the behaviour of Mendel's particles and the chromosomes during meiosis.

In 1909, working on the fruit-fly *Drosophila melanogaster*, Thomas Hunt Morgan found some odd patterns of inheritance. They gave the first proof that genes had a physical existence and led ultimately to modern molecular biology.

For Mendel it made no difference which was the male and which the female parent in a particular cross. For example, pollen from a green-seeded plant fertilizing the egg of a yellow-seeded one produced the same ratios as did 'yellow' pollen fertilizing 'green' eggs.

In the *Drosophila* cross, though, it did make a difference which parent (male or female) had which phenotype. Morgan used a pure line in which every individual had white eyes rather than red. When white-eyed males were crossed with red-eyed females from a different line, and the offspring of that cross mated among themselves, the results were not very different from those obtained by Mendel. All the F_1 flies had red eyes; and there was a ratio of three red to one white in the F_2. There was, however, an oddity; all the F_2 white-eyed flies were male (Figure 1.4a).

The oddity became more marked in the **reciprocal cross**: white-eyed females with red-eyed males. This time, in the F_1, half the flies had red eyes and half white; and, once again, all the white-eyed flies were male. When these flies were crossed, an F_2 ratio of half white- and half red-eyed flies appeared. This time, though, there were equal numbers of both male and female red- and white-eyed individuals (Figure 1.4b).

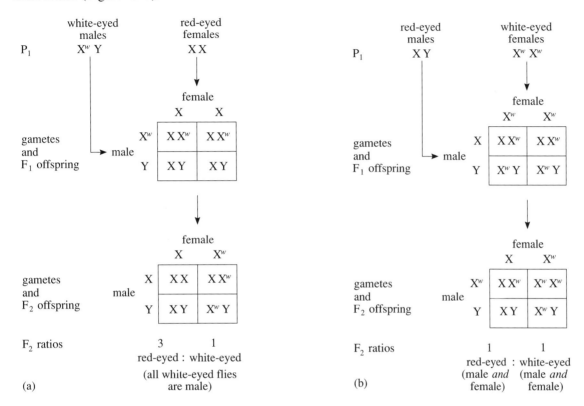

Morgan noticed that the pattern of inheritance of eye colour followed that of one particular chromosome, the X (which is one of a specialized pair of chromosomes, the sex chromosomes). In mammals and most insects, males and females are distinct in that the males have one large (the X) and one small (the Y) member of this chromosome pair while females have two more or less identical copies of the X chromosome; that is, males are XY, females XX.

Figure 1.4 Morgan's experiments on the inheritance of eye colour in *Drosophila*.

Morgan's result was explained. The pattern of inheritance of the white-eye allele followed that of the X chromosome. In males, any allele on the single X shows its effects as there is no second X chromosome bearing an allele to mask it. These males are referred to as **hemizygous**. In the first cross, the hemizygous F_2 males have white eyes, while the heterozygous females are *wild-type* (that is, the type found most frequently in natural populations). The locus for eye colour must actually be on the X chromosome. Thus from these experiments Morgan was able to see the first real fit between a genetic model based on patterns of inheritance and a physical structure in the cell.

▷ What is the phenotype and genotype of the daughters of a cross between homozygous red-eyed females and white-eyed males? What would be the phenotype and genotype of the offspring if these daughters were then mated with their fathers?

▶ All the daughters are heterozygotes, with red eyes. Their own offspring will, on mating to their fathers, contain equal numbers of red-eyed males, red-eyed females, white-eyed males and white-eyed females. This is a **backcross**.

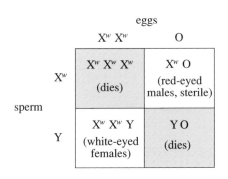

Figure 1.5 The explanation of Morgan's unexpected result in his experiments on the inheritance of eye colour in *Drosophila*: rare non-disjunction of the X chromosomes in the female parent.

The final proof that genes were on chromosomes came from taking advantage of an odd quirk in Morgan's stocks. Very occasionally, when white-eyed females were crossed with red-eyed males, there was an unexpected result. About one offspring in 1 000 was either a white-eyed female or a red-eyed male. It soon turned out that these flies were chromosomally abnormal: during the formation of their mother's eggs there had been an error. Instead of one X chromosome passing to each egg (as in normal meiosis), some eggs had two Xs and others had none. They were the result of **non-disjunction** of the X chromosomes. When these eggs are fertilized with normal sperm, some unusual zygotes are produced (Figure 1.5). Flies with three Xs or just a single Y all die, leaving four classes; typical red-eyed males and white-eyed males produced by normal gametes (as in Figure 1.4b), females with two X chromosomes and a Y, and males with but a single X.

The genetically anomalous flies are also chromosomally abnormal. The fit between the physical structure – the chromosome – and a hypothetical particle inferred from the results of breeding experiments gave the final proof that genes are indeed located on chromosomes. As Morgan put it, in a phrase famous among (but often ignored by) geneticists: 'Treasure your exceptions!'

1.4.3 Mendelian inheritance in humans

We now know that Mendel's laws also apply to humans. All his inferences from peas apply to ourselves. Human genes are particles that segregate: except in abnormal circustances, two alleles at a locus are never present in the same gamete but always separate and pass to different gametes. What is more, they assort independently: alleles move to gametes without reference to the behaviour of alleles at other loci.

The consequences are beautifully simple. If a man and a woman have identical alleles, *A*, at a particular locus, they can only produce gametes of type *A* and can only have children of type *A A*. But if, for example, the genotype of the father is *A A* and that of the mother *A a* then although the father can only produce *A* gametes, half the mother's gametes will be *A* and the rest *a*.

▷ What are the possible genotypes of the children in a cross between an *A A* father and an *a a* mother?

▶ The father produces only *A*-bearing sperm, the mother *a*-bearing eggs. All the children will therefore be *A a* heterozygotes.

▷ What if both parents are *A a*?

▶ The father produces both *A*- and *a*-bearing sperm; the mother produces *A*- and *a*-bearing eggs. When sperm and egg come together, there are three possible classes among the offspring, *A A*, *A a* and *a a*.

▷ What is the probability of a child from this mating being *a a*?

▶ Half the father's sperm are *A*, and half are *a*. In the same way, half the mother's eggs are *A* and half *a*. The chance of any single sperm or egg carrying the *a* allele is 0.5, and the chance of an *a* sperm fertilizing an *a* egg is $0.5 \times 0.5 = 0.25$.

1.4.4   Inferring patterns of inheritance from pedigrees

Comparing relatives is central to human genetics. As it is obviously impossible to carry out planned breeding experiments, one must depend on family trees or *pedigrees* with the hope that 'informative' matings have happened by chance.

The first human pedigrees were published remarkably soon after Mendel's laws were rediscovered. One showed the inheritance of brachydactyly (short fingers) in a Norwegian village (Figure 1.6). This is a classic example of dominant inheritance: those with the condition always descend from an affected parent and among the children of an affected parent there are always (given the statistical uncertainties due to limited family size) half affected and half unaffected. No unaffected person who marries someone similarly unaffected has affected children: the gene has disappeared from their family line. With the exception of rare genetic accidents, or mutations, discussed further in Chapter 4, this is characteristic of dominant inheritance.

▷   What are the probable genotypes of individuals III-3 and IV-7; and what can we say with some certainty about the parents of individual I-1?

▶   III-3 is a homozygote for the normal allele (which is recessive to that causing brachydactyly). He shows no symptoms of the disease, and following his marriage to an affected female, half his children are affected. IV-7 has short fingers (as have her father and three of her seven children). She is a heterozygote for the dominant allele producing brachydactyly. No information has been preserved about the parents of individual I-1; but given that she is affected it is very likely that one of them had short fingers. There is, however, a slight possibility that they had a normal phenotype, but that there was a mutation at this locus in the cells producing sperm or egg.

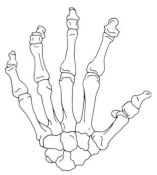

brachydactyly (short fingers)

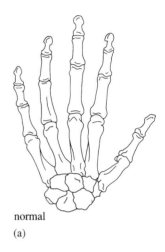

normal
(a)

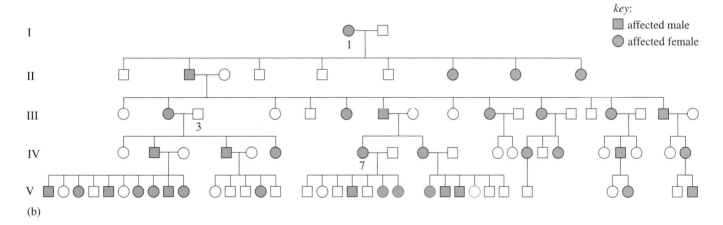

key:
■ affected male
● affected female

(b)

Figure 1.7 is an unusually lengthy pedigree, stretching over nine generations, of four Swedish families affected in recent generations by an inherited condition that causes mental defect, spastic limbs and scaly red skin. Because so many people are involved, the convention is followed of not showing all of them; in particular, many marriage partners are missed off if they do not show the condition and are not closely related to their spouse. There is also the convention that four normal sons in a row (for example) are shown as a square with the number 4 within it (there are several examples in generation VII).

Figure 1.6 Brachydactyly (short fingers), an inherited condition in humans that is caused by a dominant allele. (a) The brachydactyly phenotype. (b) Pedigree for a family with brachydactyly. In this and other pedigrees you will meet in this book, upper case Roman numerals down the left-hand side denote the generations.

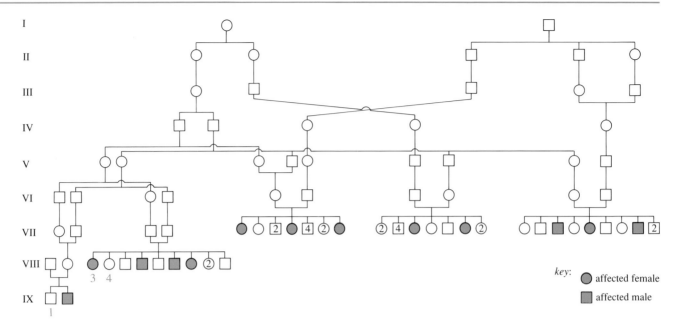

key:
⬤ affected female
⬛ affected male

Figure 1.7 Pedigree of four Swedish families, derived from two original couples born in the early 18th century, affected by a recessive mutation which, in the homozygous state, leads to mental defect, spastic limbs and scaly red skin.

▷ What is the probable pattern of inheritance and the probable genotype of individuals VIII-3 and VIII-4?

▶ This is autosomal recessive inheritance (**autosomal** = not on the sex chromosomes). Note that, quite frequently, affected children are born to normal parents. Also, within a family containing affected children – and given the statistical uncertainties associated with the relatively small size of human families – there is a ratio of three normal to one affected child.

Given that the condition is an autosomal recessive, VIII-3 is homozygous for the abnormal allele. VIII-4, though, could be either be homozygous for the normal allele or a heterozygote. There is no way of telling without looking at a sufficient number of her descendants.

▷ What is the probable genotype of everyone in generations I–VI?

▶ They are all heterozygotes (in fact, they represent that small sample of the whole enormous kindred that must be carriers of the allele; scores of others have simply been missed off). They are defined as being heterozygotes as they have affected descendants, sometimes many generations later. As each of them married an unrelated normal person (who is highly unlikely to have carried this particular very rare allele) the condition did not show itself (homozygotes did not appear) until generation VII.

▷ If VIII-4 and IX-1 marry, what is the chance of their first child being affected by the disease?

▶ VIII-4 and IX-1 both have affected brothers and sisters. This shows that their parents must both be heterozygotes. The chance of VIII-4 being a heterozygote is hence 1 in 2. The same is true of IX-1. The chance of *both* of them being heterozygotes is thus 1 in 2 multiplied by 1 in 2 – that is, 1 in 4.

The chance of a child of two heterozygotes being affected is 1 in 4. Therefore, the chance of the first (or any) child of VIII-4 and IX-1 having two copies of the recessive allele, and thus itself being affected, is 1 in 4 multiplied by 1 in 4, which is 1 in 16.

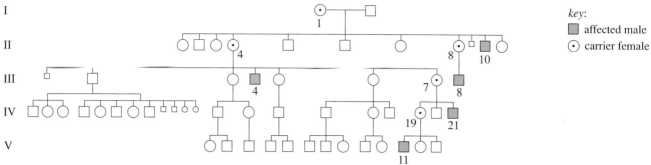

Figure 1.8 A pedigree for Duchenne muscular dystrophy, an X-linked recessive trait which usually causes the death of affected males before they are old enough to have children. (The small symbols denote stillbirths.)

There are also plenty of examples of sex-linked (or, to be more precise, X-linked) inheritance in humans. Figure 1.8 shows the patterns of transmission of the muscle-wasting disease, Duchenne muscular dystrophy. Such pedigrees have several distinctive features.

1   Only males show symptoms of the disease.

▷   Does this in itself prove that a character is sex-linked?

▶   No – there are plenty of characteristics that show themselves much more frequently in males or in females, even though they are autosomal. This is true for example, for the rare inherited form of breast cancer. Sometimes the effect of sex is subtle. One particular gene causes sex-limited baldness among males and is transmitted as an *autosomal* dominant. Females rarely go bald as they do not have the hormonal milieu in which the allele can show its effects.

2   In this pedigree, unaffected men never have affected children.

▷   Is this always true for sex-linked characters?

▶   No. For example, a man might marry a woman who is herself a heterozygote, in which case half the daughters may be affected.

3   All the sons of women who have the disease (rare though such women are: there are none in the pedigree above) themselves have the disease.

▷   What is the probable genotype of a daughter of an affected mother?

▶   She will be a heterozygote for the character; and half her own daughters are likely to be heterozygotes too.

The existence of distinct characters, controlled by separate loci, on the X chromosome introduces the most important revision of simple Mendelism. If two loci are both on the X, and the X is the physical unit of inheritance which can be seen to be passed down from generation to generation, then, for the loci on this chromosome (or, for that matter, on any other) Mendel's second law cannot hold. They do not show independent assortment. Instead, they tend to be inherited together – they are **linked**. This is true, for example, of colour-blindness and muscular dystrophy in humans. As we shall see (Chapter 2) linkage is not absolute, but it means that Mendel's view of the genes as a cloud of independent particles must often be wrong.

Rarer than X-linked genes are those that are Y-linked. One gene which can confidently be ascribed to the Y is the one that directs the developing embryo into the path of maleness.

▷ What would be the pattern of transmission of this 'testis-determining factor' gene in a family?

▶ A male would pass it to all his sons (as all have a copy of his Y chromosome); daughters would never show the trait.

1.4.5 Mendelian inheritance is simple, genetic counselling less so

Armed with a knowledge of simple Mendelian genetics, it is sometimes possible to give genetic advice to prospective parents.

▷ If a phenotypically normal couple have had two children, one with and one without the recessive illness Tay–Sachs disease, what is the chance of their next child being affected? (Tay–Sachs disease is an inherited degeneration of the nervous systen which leads to early death.)

▶ One in four: as one child was affected, then both parents must be heterozygotes. Every time they have a child, there is a one in two chance of sperm carrying the abnormal allele, and exactly the same chance for eggs. The chance of sperm and egg *both* having the abnormal allele is hence one in four. It is not, of course, the case that having had one affected child somehow reduces the chance of the next having the disease (as some parents believe). Think of spinning a penny twice: the chance of getting heads on the second spin is not altered by the fact that heads also came up first time round.

This logic is beguiling in its simplicity. It certainly works well in many cases where parents come to a genetic counsellor for advice. However, in practice, genetic counselling is more complicated than this. Sometimes it has to resort to the traditional pragmatic approach of medicine: predicting the likelihood of recurrence of an inborn condition in a family only from empirical information on how frequently the affliction has reappeared in other families (which may be rarely).

There are many reasons why genetic counselling is often a complicated business. The patterns of inheritance of many inborn diseases have been established from pedigrees. However, some are so rare that not enough cases are known to allow this to be done with confidence. There are other difficulties in family studies, too. Particularly in the modern world most families are too small to provide enough data to work out genetic ratios (consider, for example, the size of the Norwegian kinships shown in Figure 1.6). Certain groups – Mormons in the United States, and some Bangladeshis in Britain – still have large numbers of children, and these peoples are particularly useful to geneticists. However, many diseases appear in families with so few children that the pattern of inheritance cannot safely be inferred.

Other inborn diseases involve not one but many genes acting simultaneously; yet others may not involve specific genes at all but have arisen from some unknown problem during pregnancy. Sometimes, two diseases with very similar phenotypes may have different genetics. Brachydactyly (the short-finger phenotype described earlier) is an autosomal dominant disorder. However, Albright hereditary osteodystrophy – which has a similar phenotype – appears to be an X-linked dominant condition. Clearly, this is important to genetic counselling since an X-linked allele cannot be transmitted from father to son but an autosomal one can.

Sometimes, one form of a disease is strongly familial (inherited) while another – apparently very similar – seems to be scarcely inherited at all. This is true, for example, for breast cancer in which only about one case in twenty is clearly genetic in origin.

Diagnosis may also be a problem – for example, although manic depression is generally agreed to have some hereditary component, there is much disagreement between schools of psychiatrists as to just what criteria must be fulfilled if a patient is diagnosed as having the disease. In manic depression, as in many other diseases, the environment plays an important part and this too can confuse the interpretation of pedigrees.

There is also the important problem that some children are afflicted with inborn disease because of a new genetic error in the sperm or eggs of their parents or grandparents. This, too, confuses the advice that might be given to parents planning whether or not to have another child. Molecular biology has revolutionized genetic counselling, at least for single-gene disorders: as we shall see, it is now possible to identify many abnormal alleles in heterozygous condition or to diagnose an affected fetus before birth. However, it has also provided new problems. It became clear that many genetic diseases that once appeared simple are in fact *heterogeneous* in that the malfunctioning of many different genes can result in similar phenotypic conditions. For certain conditions, only some of the abnormal alleles can be detected with standard tests. Counselling, like genetics itself, is a lot more complicated in practice than might seem reasonable given the beautiful simplicity of Mendel's theory.

1.4.6   Genes can be studied in populations as well as in families

It is a truism of biology that a chicken is simply an egg's way of making another egg. In other words, it is often necessary to separate the fate of genes from that of those who carry them. *Population genetics* is the study of genes in groups of people (or other creatures), rather than just in single carriers or their relatives. It is based on exactly the same simple logic as that used by Mendel.

Among some groups of Ashkenazi Jews, about one child in 1 000 is born with the recessive inherited condition Tay–Sachs disease. That is, one in every 1 000 matings in the population as a whole must have been between two individuals who were heterozygotes, and both sperm and egg must have carried a single copy of the gene.

▷   What, in the population as a whole, is the chance of a single sperm (or a single egg) carrying the recessive Tay–Sachs allele?

▶   If the chances of two gametes drawn at random *both* possessing the allele is 1 in 1 000, then simple laws of probability tell us that the chance of one sperm or one egg taken at random from the population as a whole carrying the allele is the square root of this – that is, one in about 32. The number of heterozygous individuals carrying a single copy of a rare recessive allele without manifesting its effects hence far outweighs the number of affected children, born with two copies.

The relationship between the frequencies of homozygotes for the dominant allele, heterozygotes and recessive homozygotes is given by a simple formula, the Hardy–Weinberg equilibrium.

Box 1.1 The Hardy–Weinberg equilibrium

A cross between two heterozygotes for the Tay–Sachs allele can be summarized as shown below, where t is the Tay–Sachs allele and T is the normal, dominant allele:

	eggs	
	$0.5\ T$	$0.5\ t$
sperm $0.5\ T$	$0.25\ TT$	$0.25\ Tt$
$0.5\ t$	$0.25\ Tt$	$0.25\ tt$

In other words, there is a ratio of one TT homozygote to two Tt heterozygotes to one tt homozygote among the offspring of this couple.

Just the same logic comes into play when considering the fate of the gene in the *population*. We can ignore the fact that the two alleles involved are carried by individuals, and consider them just during their brief existence as sperm or eggs. If the frequency of the recessive allele t among this group of Ashkenazim is 0.032 (3.2%), then that of the normal allele T must be 0.968 (96.8%). We can redraw the simple diagram above as:

	eggs	
	$0.968\ T$	$0.032\ t$
sperm $0.968\ T$	$0.937\ TT$	$0.31\ Tt$
$0.032\ t$	$0.31\ Tt$	$0.001\ tt$

In the next generation of the population as a whole, then, 93.7% of the offspring will be TT, 6.2% Tt and 0.1% tt. Notice how much more abundant the t allele is among phenotypically unaffected Tt heterozygotes than it is among children born with the disease – in this case, the frequency of Tt heterozygotes is more than 60 times as high as that of tt homozygotes; that is (given that heterozygotes have a single copy of the t allele), only about one t allele in 30 in the population as a whole manifests its effects in an individual who is born with two copies.

▷ What is the relevance of this to those who say that it is our eugenic duty to prevent those with two copies of an allele for a recessive disease from reproducing?

▶ Such a programme would be very ineffective: for rare diseases, selection against heterozygotes is the only way of altering gene frequencies over a reasonable period.

Exactly the same logic applies whatever the frequency of T and t. Putting it into algebraic terms, we can say that:

The frequency of T is p.

The frequency of t is q.

As there are only two alleles, the sum of p and q must be 1.

Drawing the diagram again, we get:

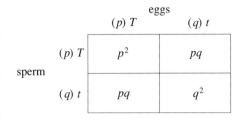

	eggs	
	$(p)\ T$	$(q)\ t$
sperm $(p)\ T$	p^2	pq
$(q)\ t$	pq	q^2

In other words, the frequencies of the three genotypes are:

TT	Tt	tt
p^2	$2pq$	q^2

Any frequency between 0 and 1 can be substituted for p and q. This **Hardy–Weinberg equilibrium**, as it is known, is very useful in population genetics. Although it makes certain assumptions (for example, that there is not a constant influx of new Tay–Sachs cases from elsewhere) it can be used in many ways. Thus, it is possible to calculate the numbers of heterozygotes in a population ($2pq$) just from information on the number of affected children (q^2).

Around 3000 'proven' and 2500 'probable' monogenic diseases (i.e. diseases caused by a single gene defect) are now known. Many are controlled by dominant alleles. However, at least half are under the control of recessive alleles. Because a person must inherit two copies of a recessive allele to show the symptoms of a disease this means that even if the number of people born with the condition is small the gene itself may be quite common. Of course, the frequencies of different recessive alleles vary greatly from place to place: no other group has as many copies of the Tay–Sachs gene as do Ashkenazim. Nevertheless, the Hardy–Weinberg law in consort with the information on the incidence of genetic disease suggests that most people must, quite without knowing it, be carriers of single copies of a harmful recessive allele at one locus or another. One estimate is that, on average, everyone carries about two damaging recessive alleles in heterozygous condition. As we shall see (Chapter 4), at one time the suggestion that the human race possessed much hidden genetic damage – as it certainly does – caused great scientific and public concern. There has been a remarkable shift in opinion since then; why it is hard to say.

1.4.7   *Everything that is inherited is not genetic*

Pedigrees are so alluring that it is easy to forget just what they represent: patterns of inheritance of *characters*, not *genes*. One can make impressive pedigrees of the inheritance of country houses, or criminal records, or university degrees: but inheritance in itself does not prove that a character is genetic. Most human attributes cannot be categorized into distinct classes as was possible for Mendel with his peas or Morgan with his flies. Galton's own approach to human inheritance was a **biometric** one. In his work on 'Heredity, Talent and Character' he measured, as accurately as possible, traits such as stature or intelligence in individuals of known relationship – brothers and sisters, twins, or parents and their children. He then used statistical methods to compare them.

The measurements were *continuous*: people might be graded on some sort of scale, but they could not sensibly be sorted into distinct classes. Galton himself saw that such traits – **metric** traits, which are measured rather than counted – were influenced by both nature and nurture; but he was strongly of the opinion that they were very much under the influence of heredity. In fact, the simple observation that a trait runs in families in itself says very little. There was a reaction among geneticists in the earlier part of the 20th century against the whole Galtonian idea: for many years it seemed that human inheritance could not be explained in the simple terms which applied to peas. Part of this arose from the publication of absurd pedigrees which claimed to explain human behaviour in simple genetic terms: there was, for example, supposed to be an allele for 'drapetomania' – hereditary running away among slaves.

There was a bitter conflict between 'Mendelians' and 'biometricians' about the nature of genetics. It was – at least partially – solved when it was pointed out that correlations between relatives for metric characters could be explained if each was the sum of the effects of many different genes.

Mendel dealt with phenotypes, not genes. He had no idea just why one pea was yellow and another green. He had, though, what every scientist needs to be recognized as a genius – good luck; the good luck (tempered with good judgement) to work on simple characters in a simple creature. His ratios themselves were beguiling in their clarity. Most of the examples of human inheritance mentioned so far are straightforward in that they represent genetic disease; traits which distinguish those who have them from those who do not. Some quite normal traits (blood groups, for example) follow the same simple rules. People, though, are more complicated. They are taller or shorter, fatter or thinner, more or less intelligent or more or less likely to contract cancer. Certainly, such characters run in families, but even the most ardent hereditarian does not claim that they follow Mendel's ratios.

It has taken more than a century to realize just how lucky Mendel was. The world of modern genetics is full of anomalies. The nature of genes – and, even more, of phenotypes – has turned out to be far more complicated than it once seemed reasonable to predict. The rest of this book deals with some of the complications and fascinations of genetics since Mendel. Before you read it, we should add a 'health warning': for reasons of convenience we ourselves (in spite of the dire warning implicit in the preceding paragraphs) often refer to a gene 'for' a particular character. This is a shorthand, of course. Any character, even the seemingly most simple, like eye colour, is the result of a coordinated developmental sequence involving thousands of gene products. A gene 'for' a particular eye colour really means a gene *in the absence of which* a different eye colour results. The 'character' is the result of all the rest of the genes *minus* the one for the character, working together. All genes work in the context of the environment in which they are placed, and this environment includes all the other genes and their products. We also sometimes talk of 'genetic' disease. Again, this is little more than a convenient shorthand. Many inborn diseases – some of which follow conventional patterns of Mendelian inheritance – can be successfully treated with drugs or surgery. In other words, a 'genetic' disease is perfectly susceptible to 'environmental' modification. In the same way, a disease that might at first sight seem to be due solely to some environmental agent – the malaria parasite, for example – often turns out to depend for its effects on the genotype of its victim. It is tedious to keep repeating the dangers implicit in an uncritical use of language, but it is worth reminding yourself of them at frequent intervals. To do so should do something towards reducing the triumphalism which all too often infects human genetics.

1.4.8   Modifying Mendel: a molecular microcosm

To give a flavour of the subject as a whole, and as a hint of the intricacy of real life, we spend the remainder of this introductory chapter looking at the haemoglobins. These are the proteins – and the genes – that, because of the existence of many mutations resulting in specific and easily studied blood disorders, began the task of turning humans into the genetical equivalent of fruit-flies. Biochemistry, too, began with haemoglobin: crystalline haemoglobin was prepared as early as 1849.

Inherited errors in haemoglobin structure provide the single greatest genetic burden inflicting the human race. Haemoglobin has become the model system of human genetics, giving the first hint of what has become the new science of molecular medicine and illustrating, better than anything else, how what seemed at first sight a straightforward gene doing a simple job has turned into an enormously complex system about which a great deal remains to be discovered.

1.5   Normal and abnormal haemoglobin

The major role of red blood cells is to carry oxygen from the lungs to the tissues, and to take part in the transfer of carbon dioxide in the other direction. To this end they are remarkably simple objects, filled with the four-subunit protein, haemoglobin, and lacking a nucleus (which is lost early in their development). Figure 1.9 shows the structure of a haemoglobin molecule.

Haemoglobin could have been invented for the convenience of biologists. It is abundant, readily available, and comes in pure form. The protein is small and has a shape that is easy to study. Even better, the various globins that appear in different parts of the body and at different times during development are controlled by a series of related gene loci whose activity alters as the embryo develops. For those whose interests extend into evolution, globin genes and their relatives are found in creatures as different as peas, frogs and mammals; and the relationship between globin abnormalities and infectious disease is the classic example of natural selection in humans, including the first of all the intensively studied genetic diseases, sickle-cell anaemia.

There are also important medical reasons for studying haemoglobin. Inborn errors in its structure are by far the commonest monogenic diseases in the world. There are already around 250 million people who carry a single copy of a damaged globin gene; and by the year 2 000 one person in 15, world-wide, will be a carrier. Sickle-cell disease, the best-known of the many hereditary anaemias due to

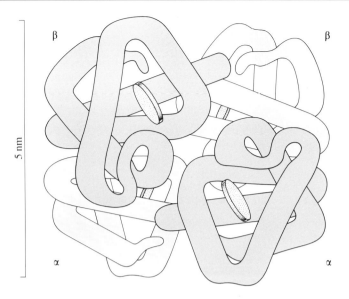

Figure 1.9 The quaternary structure of haemoglobin consisting of four subunits (globin chains). The adult form, shown here, is tetramer of two α (alpha) and two β (beta) chains. The grey discs are the oxygen-binding haem groups.

haemoglobin defects, was described as early as 1670 among the Krobo people of Ghana. It was shown to be a fault in red-cell structure as long ago as 1910. As we shall see, it represents the simplest possible genetic change: a change in one nucleotide in the genetic message. This alteration in the DNA code changes a crucial amino acid in the haemoglobin. In turn, this alters the oxygen-binding properties of the molecule, causing the red cells to assume a 'sickled' shape (Figure 1.10) in conditions where oxygen concentration is low and leading to a range of painful and debilitating symptoms.

The extraordinary abundance of this apparently harmful gene is due to **heterozygote advantage**. Perhaps because the malaria parasite is unable to thrive within cells containing some sickle-cell haemoglobin, carriers are more resistant to malaria than are individuals with two copies of the normal allele. This advantage is enough to counterbalance the severe symptoms suffered by those with two copies of the sickle-cell allele itself. This is a classic example of natural selection in action. Ten million cases of malaria are reported to the World Health Organization each year and 90% of the population of sub-Saharan Africa lives in places where the disease is endemic.

Sickle-cell was the first 'molecular disease' to be identified – a disorder whose symptoms can be related to a single base-change in DNA. Because it is so common, it was the target of one of the first genetic screening programmes. Because of sickle-cell's historical role in malaria protection it is an illness that mainly affects those of African ancestry. One African in 25 and one black American in 100 has sickle-cell anaemia. The Hardy–Weinberg equilibrium predicts that the number of carriers of a single copy of the gene will be much higher than is that of those affected by the disease. In fact, more than a third of all Africans are heterozygotes. It is worth remembering, though, that because of the independent assortment of the β-globin locus and those for skin colour, there are plenty of white-skinned people with one or more African ancestors who have the sickle-cell gene. For example, in the town of Coruche in southern Portugal there is a focus of sickle-cell disease descending from the Africans who were taken there by the earliest Portuguese explorers. Other Iberian towns have no sickle-cell.

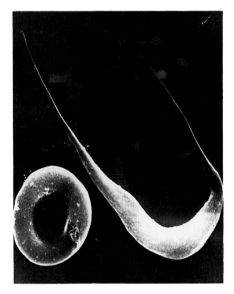

Figure 1.10 A sickled and a normal red blood cell.

Now we know that there are many other inherited abnormalities of haemoglobin. These, too, have given new insights into human inheritance and these too reach high frequency in certain parts of the world.

Haemoglobin may have a straightforward job, but it can go wrong in complicated and surprising ways. The molecule and its errors are a microcosm of how the simplicity of Mendelism is clouded when a phenotype (or a gene) is studied in sufficient detail. What seems at first sight an uncomplicated piece of machinery for transporting oxygen around the body is in fact an intricate apparatus with several parts which might, an engineer could argue, be unnecessarily elaborate. The same symptoms may result from a whole series of distinct genetic errors; and the same genetic fault can produce very different symptoms in different carriers.

To end this introductory chapter we will discuss how Mendelism must be modified to accommodate the complexities of the real world of human genetics, as manifest in the study of haemoglobin disorders. To understand them demands a series of new explanations (which might have baffled Mendel). How genes are studied and the details of how they work will be postponed until later chapters: but haemoglobin, better than anything else, illustrates the revolution which has taken place in our views of inheritance since the advent of the new genetics – which is rapidly becoming a branch of chemistry rather than biology.

Nearly all organisms need oxygen, and large ones need a transporter to move it from the outside world to their tissues. In mammals, the transporter molecule is haemoglobin. Two pairs of protein chains (the globins) bind a central *haem* group, which contains iron and gives blood its red colour. The pattern of binding changes on deoxygenation, which is why venous blood is blueish compared to the bright red of that in the arteries. As you know from Book 1, Chapter 2, normal haemoglobin has the ability to bind oxygen, four oxygen molecules to each haemoglobin molecule (one to each globin subunit). The binding of a single oxygen increases its affinity for further oxygens. Many abnormal haemoglobins lose this property and are hence less effective at doing their job. In the tissues, where there is not much oxygen but a lot of carbon dioxide, in the form of bicarbonate ions, haemoglobin loses oxygen and, in addition, binds carbon dioxide. In the lungs, the process is reversed. As a result, haemoglobin also plays an important part in transporting carbon dioxide.

1.5.1 One simple task demands many complex genes

The job of carrying oxygen around the body might seem straightforward, but it is not. Haemoglobin is only part of a chain, many of whose links are under genetic control (and many of which can go wrong). The molecule itself is quite complicated. There are several distinct haemoglobins, each made up of two pairs of dissimilar protein chains (Figure 1.9).

Nearly all the haemoglobin in adult blood is in a form known as HbA, with two α- and two β-globin chains ($\alpha_2\beta_2$, for short). About 3% of the total, though, is in a slightly different arrangement, HbA_2, in which the β chains are replaced with delta chains, δ ($\alpha_2\delta_2$). The fetus has its own needs for oxygen. It cannot, of course, obtain it directly from the air but must force its mother to give up some of her own supply. To this end, the developing embryo has, for most of its life, another form of the molecule, in which the β chains are replaced with gamma chains, γ. This is fetal haemoglobin or HbF ($\alpha_2\gamma_2$) and traces of this form are present even in adults. Fetal haemoglobin has a higher oxygen affinity than does the adult form; this means, of course, that the fetus can draw oxygen from its mother's blood. The very early embryo has two further versions, involving two additional chains, ε (epsilon) and ζ (zeta) in various combinations with each other and with α and γ. (See Figure 1.12.)

Myoglobin (a close relative of haemoglobin which is present in muscle) can wrest oxygen from haemoglobin when demand in the muscles is very high.

Each haemoglobin has its own shape, but in every one the protein chains are folded around each other (and around the iron-containing haem group) to give a globular outline with a groove which binds oxygen when required. Although only about one amino acid in five is identical in myoglobin, α-globin and β-globin, their three-dimensional shape (the tertiary structure) is similar.

The gain and loss of oxygen involves a change in the shape of haemoglobin as the chains slide over each other (Figure 1.11). When oxygenated, the molecule is said to be 'relaxed'; when not so, it is 'tense'. Much is known about how the chains interact with each other and with oxygen. Some abnormal haemoglobins lack essential binding sites between chains, and this alters their ability to pick up oxygen.

There are two distinct groups of globin chains. One, the α-globin group, includes α and ζ; members of this group have 141 amino acids. The other, the β-globin group, includes β, δ, ε and γ, and these chains have 146 residues. Although related, the two groups presumably diverged in the distant evolutionary past.

1.5.2 Much of the genetic message is interrupted

The globin genes make up a **gene family**: a group of genes with related function and a common evolutionary ancestor. Two separate arrays of genes are involved, corresponding to the globin groups described above. The α-globin cluster is on chromosome 16, and the β-globin group on chromosome 11 (Figure 1.12). Their products cooperate to form haemoglobin. Each of the globin chains is coded for by a separate stretch of DNA.

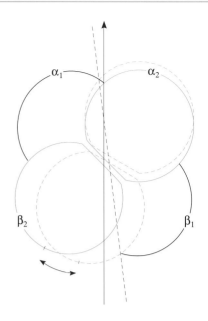

Figure 1.11 Schematic diagram of the quaternary structure of haemoglobin showing the relative orientation of the chains during oxygenation (dashed outlines).

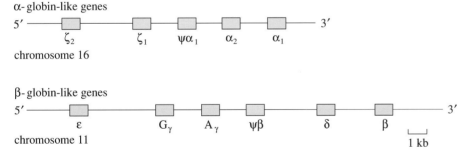

Figure 1.12 The arrangement of the α-globin-like genes on chromosome 16 and the β-globin-like genes on chromosome 11. (ψ (psi) indicates a non-functional *pseudogene.*)

Using the technology described at greater length in Chapter 2, the position of each of the globin genes in the genome has been established. Although the two gene families are unlinked (and hence show independent assortment), the members of each family are very close together on their shared chromosome. As a result, they are transmitted together as a group from one generation to the next. Occasionally, the linkage is disrupted by recombination (see Chapter 2), but in general the length of DNA encompassing each gene family is inherited as a unit.

The details of the arrangement of the globin genes within each family gave a first hint about the complicated (and in many ways startling) structure of the genome. Underlying a simple and important part of the body's machinery – the protein that transports oxygen – is a complicated genetic mechanism. Before molecular biology, most of this complexity was hidden from view. Biologists are now able to go straight to the genetic message itself by examining DNA directly (rather than being limited to trying to infer its contents from what it produces, as Mendel was).

Studies of the sequence of the DNA in each of the globin gene families showed that – quite unknown to those able to see only the products of genes – other, related, DNA segments were present. The order of bases in each one was quite close to that of the working genes in each group, but none was capable of making a working protein. These **pseudogenes** are probably relics of functioning genes which have been destroyed during evolution by the accumulation of mutations.

The structure of the working globin genes is also full of surprises. The information within the genes themselves is interrupted by non-coding sections, and in between the globin genes there are long sections of DNA that appear to code for nothing. At the molecular level, genes are far from being the neat particles hinted at by Mendel.

1.5.3 Most genes are switched off most of the time

Nobody knows how many functional genes there are: one guess is that it may take between 50 000 and 100 000 to make a human being. All the cells of a particular organism have all its genes, but, in most tissues, most of them are switched off most of the time. One of the most important (and least understood) aspects of genetics is the control of just what genes are activated where. Cells differ greatly in their needs: some, such as nerve cells, never divide, while others undergo mitosis every few hours. Some tissues – the pituitary gland, for instance – produce many complicated products. Red blood cells are at the other extreme: their major protein product is haemoglobin and they use only a few other genes (those concerned with the cell's 'housekeeping' functions) in addition to the haemoglobin genes. In most mammals (but not in birds and other creatures) they can afford to lose the nucleus and the genes it contains altogether, once they become mature.

Few cells go quite this far in repressing unwanted genes; but all control their expression to a greater or lesser extent. Every cell contains all the genetic information transmitted to sperm and egg – far more hereditary particles than is manifest in its own phenotype. Often, the pattern of gene activation depends on circumstances. It may, for example, alter during development (Book 1, Chapter 1). This is the case for the haemoglobin genes, as the demand for oxygen and the sources from which it is obtained change greatly from embryo to adult. Different members of each globin gene family, each producing a protein with differing oxygen affinity, are switched on to cope.

In the very early embryo, specialized haemoglobins made up of α, ϵ, γ and ζ chains in various combinations are made in the yolk sac. After about eight weeks, the liver begins to make fetal haemoglobin (α and γ chains), and this is slowly overtaken by the production of both HbF and HbA in the bone marrow. HbF continues to be made for the first few months after birth. Figure 1.13 shows the shift in activity of the various members of the two gene families during embryonic development and beyond. It is not known just how the move from one to the other is controlled – although it would be interesting to find out, as it might open the possibility of persuading the γ-globin genes of people with inherited defects of, say, the β-globin locus to remain active during adult life. There is discussion of the possibility of treating a deficiency in one gene by persuading another one to switch on in its place as so many genes are – like haemoglobins – members of gene families. However, the prospect of so doing still remains distant.

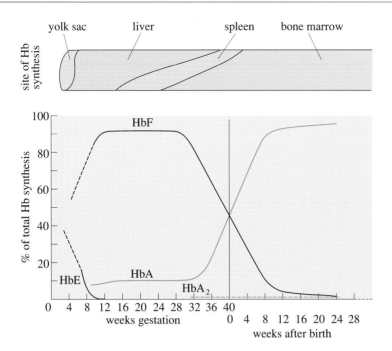

Figure 1.13 The changes in human globin chain synthesis during development. The sites of haemoglobin synthesis and red cell formation during the period shown are also indicated at the top of the figure. Types of haemoglobin: HbE = embryonic ($\alpha_2\varepsilon_2$); HbF = fetal ($\alpha_2\gamma_2$); HbA = most abundant adult form ($\alpha_2\beta_2$); HbA$_2$ = minor adult form ($\alpha_2\delta_2$).

1.5.4   Gene replication is inaccurate

The laws of inheritance depend on stability. DNA is, indeed, generally unchanging: but its constancy is not absolute. Quite often there are errors – mutations – during replication which produce slight changes in the genetic message from generation to generation or in the dividing cells of a single body. A few mutations cause inherited disease; the majority have no discernible effect at all, and a very few produce a DNA that functions more effectively than what went before. The last class is the raw material of evolutionary progress.

Mutation was once thought to be a rare and generally deplorable event. Another of the surprises brought by molecular genetics was to discover that mutation is common – and that some of the body's own systems (most notably, the immune system; Book 3, Chapter 4) actually depend for their function on repeated mutation. Errors in copying the inherited message have taken place so frequently in the past that no two individuals have exactly the same sequence of DNA in the region around the globin genes. Such differences all arise from mutations – either in parental sperm or egg or, more often, in generations long ago. What is more, older people may have in their bodies haemoglobin with a genotype different from that received from their own parents (these variants include sickle-cell haemoglobin). This is because of **somatic mutation**: errors in copying the genetic message in body cells. Old people have undergone more cell divisions and there have been more opportunities for damage of this kind.

Because there is so much diversity at the DNA level from person to person, and because members of each globin gene family are so closely linked together on a short section of chromosome, groups of mutations tend to travel together down the generations. For every individual they represent a sort of genetic 'surname' which shows the family lineage to which they belong. There are only a limited number of such **haplotypes** in any population. Figure 1.14 shows the distribution of some of the common haplotypes, each representing a set of mutational changes which happened long ago, based on just seven variable sites (of many hundreds of possibilities) in the β-globin complex. There are just nine widespread haplotypes.

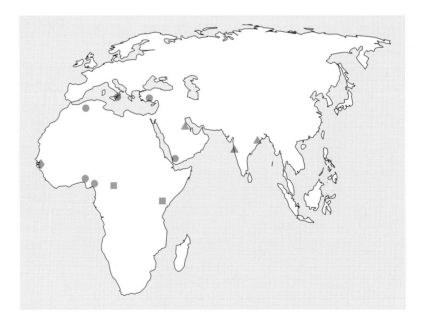

	Hc	Hd	Hd	Hc	Hc	A	B
1	+	−	−	−	−	+	+
2	−	+	+	−	−	+	+
3	−	+	−	+	+	+	−
4	−	+	−	+	+	−	+
5	+	−	−	−	+	+	−
6	−	+	+	−	−	−	+
7	+	−	−	−	−	−	+
8	−	+	−	+	−	+	−
9	−	+	−	+	+	+	+

Figure 1.14 Seven variable sites (arrowed) in the human β-globin gene complex and nine of the common haplotypes derived from them. The plus and minus signs denote whether or not the sites are cut by restriction enzymes (see Chapter 2).

▷ How many haplotypes would there be if there were free assortment, unrestricted by close linkage – and with sufficient time available to allow free recombination to exert its effects – among seven variable sites, with two alleles at each one?

▶ 128 – that is, 2^7. Close linkage greatly reduces the number of possible combinations of genotypes among loci.

Other mutational changes, though, have more drastic effects as they happen in coding regions and alter the amino acid sequence of proteins. As we have seen, the most widespread gives rise to **sickle-cell disease**. The genetic change could not be simpler. It involves a substitution of a thymine for an adenine in the

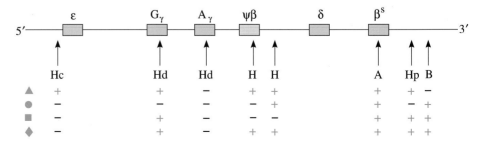

	Hc	Hd	Hd	H	H	A	Hp	B
▲	+	+	−	+	+	+	+	−
●	−	−	−	−	+	+	−	+
■	−	+	−	−	−	+	+	+
◆	−	+	−	+	+	+	+	+

Figure 1.15 Four β-globin gene haplotypes in individuals with the sickle-cell gene from different populations.

DNA coding for β-globin. This alters a crucial codon so that the corresponding amino acid – the sixth along the chain – changes from glutamic acid to valine.

The symptoms of the illness are severe. It is widespread all over the tropical world. Studies of the diversity in the DNA surrounding the crucial site show that the mutational change giving rise to the inherited error must have happened several times. In different populations, the sickle-cell mutation is associated with different haplotypes in the surrounding DNA (Figure 1.15). The simplest interpretation is that, in each separate population, the crucial change took place independently, in a long-dead individual whose DNA haplotype is preserved in the millions of sickle-cell carriers alive today. There has not been enough time since the first appearance of the mutation in each place to allow the linkage between the sickle-cell site and others nearby in the DNA to be destroyed by recombination.

1.5.5   One gene may control many characters

For Mendel, one hereditary particle did one thing. Different alleles changed the shape, the colour or the height of peas. Now, though, we know that a single genetic change can influence many aspects of the phenotype. Sickle-cell shows such **pleiotropy** particularly clearly.

As we have seen, the mutation involves just one base in the DNA. From this simple change great consequences flow (Figure 1.16).

Figure 1.16   Pleiotropic effects of the sickle-cell gene: an amino acid substitution of valine for glutamic acid in a β-globin chain.

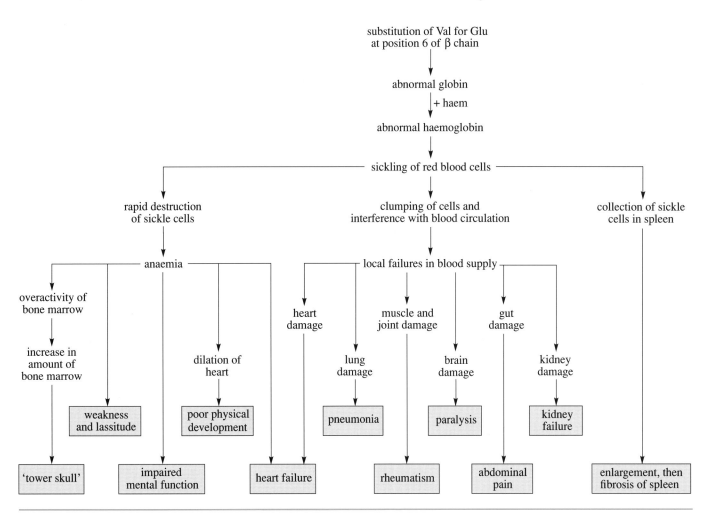

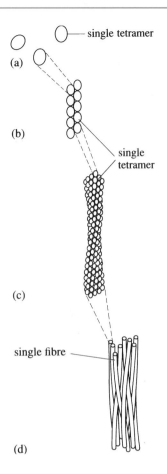

single tetramer

(a)

(b)

single
tetramer

(c)

single fibre

(d)

Figure 1.17 Polymerization of sickle-cell haemoglobin and the formation of fibres.

First, the haemoglobin itself polymerizes and forms fibres in conditions of low oxygen concentration (Figure 1.17). This deforms the cells into their characteristic 'sickle' shape. The abnormality can be reversed in the lungs, at least when the red cells are young. The switch from normal to sickled shape is quicker in an acid milieu and in parts of the body where there are high concentrations of ions. This means that the kidney and the spleen are major sites for the formation of deformed cells. After sickling, the blood becomes more viscous, and the problems begin.

If oxygen is in short supply, there is a delay before the fibres of sickle-cell haemoglobin are actually formed. The length of this delay depends on the concentration of salts in the blood (and hence on the water content inside the red cell), the amount of oxygen present, and temperature (which goes up, for example, during a fever). For much of the time, the blood returns to places where the oxygen concentration is high before the red cells have time to sickle. Only occasionally are conditions severe enough to lead to sickling and to damage. Sickle-cell disease is hence characterized by sudden painful crises rather than long-term chronic illness. These are sparked off by small changes in hydration, oxygen demand, or body temperature; any one of which can push the damaged β chain into forming fibres before the blood has time to pass through a capillary and escape to a safer place.

So fundamental are the effects of an interruption in blood flow that there are many symptoms of sickle-cell disease, which can vary in a baffling fashion from patient to patient. All reflect two basic problems: anaemia (a shortage of red cells) and the blockage of blood vessels.

Sickled cells have shorter lives than normal red cells partly because they are attacked by white blood cells which destroy them. The patient needs to replace the lost cells, but usually cannot do so quickly enough because the blood's reduced oxygen-carrying ability means that the bone marrow (the site of red cell synthesis) is itself starved of oxygen. He or she quickly becomes anaemic.

The anaemia in turn leads to a range of symptoms. Children homozygous for sickle-cell haemoglobin are pale and grow slowly. Sometimes they look slightly unusual, with a 'tower skull' and long arms and legs – characteristics due to overactivity of the bone marrow in early childhood. The spleen is also involved. The role of this organ in protecting against invaders begins to fail, and children with the disease often die from infections which to others would be harmless.

Other attributes of the phenotype emerge from the blocking of small blood vessels by collapsed red cells. Often, the children have sudden crises brought on by cold or infection. Sometimes these are very painful as patches of tissue are starved of oxygen and begin to die. If the spleen itself or the liver becomes clogged with cells, the child may die suddenly. There are chronic effects, too. The kidney is slowly devastated as the blood supply fails and one side-effect of the disease may be a raging thirst.

▷ What other organs or systems might be particularly at risk in patients with sickle-cell disease?

▶ Any organ with a high demand for oxygen, particularly if there are sudden surges in demand. For example, local failures mean that the bones may be damaged, as is the skin, resulting in painful ulcers. Even the retina can be involved, with some patients going blind. Congestion of blood vessels in the brain can produce stroke, or loss of sight, hearing or mental powers. The liver and lungs are also sometimes affected.

It seems, then, that a single DNA base change can have numerous phenotypic effects, on organs as different as the heart, the kidneys and the spleen. If we did not know so much about the molecular basis of the disease they would appear at first as quite distinct. Pleiotropy of this kind is widespread. In fact, many – perhaps most – allelic differences in working genes affect several aspects of the phenotype.

1.5.6   One gene may exist in many forms

Mendel's peas had just two options for each attribute: round or wrinkled, yellow or green and so on. Although many (perhaps most) genes exist in just one form, identical from person to person, many are **polymorphic** – different individuals have slightly different versions of that gene. Now we know that most polymorphic loci have not two but many alleles. Often, what seems to be a single variant turns out to be a complex of many distinct changes when studied with molecular technology. The inborn errors of haemoglobin show this clearly.

Sickle-cell is just one of many hereditary anaemias. All have rather similar symptoms as all lead to a reduction in the number of red blood cells. Today, of course, no-one believes that all the inherited anaemias have a common cause. Fifty years ago this seemed less obvious: although there are certainly differences in symptoms from population to population it was not impossible that every inborn failure of red cell formation reflected a diversity of effects of a single gene mutation.

In fact, scores of different haemoglobin errors are now known, each with a separate origin. Some – like sickle-cell – represent single changes in DNA sequence, but others are more complex. Hereditary anaemia (which, within a family, usually follows Mendel's laws for recessive inheritance) can, modern genetics tell us, be produced by dozens – perhaps even hundreds – of quite distinct genotypes. Just the same is true of many other inherited diseases. Again, genetics is a less straightforward business than might once have been hoped.

As well as sickle-cell haemoglobin (usually abbreviated to HbS), several other single amino acid substitutions are found in the haemoglobin chain. For example, HbC is common in West Africa. Here there is a substitution of a single amino acid at just the same point as in HbS, but this time lysine (not valine) is substituted for glutamic acid. Homozygotes have symptoms similar to, but milder than, sickle-cell disease. 'HbD' is a group of related amino acid substitutions frequent in the Punjab, again with mild symptoms. HbE (*not* the same as the embryonic form in Figure 1.13) is the commonest of all haemoglobin variants. Lysine is substituted for glutamic acid at position 26 in the β chain. It is found in Bangladesh, Burma, China, Laos and the surrounding regions. The main effect of the mutation is to slow the rate at which red cells are made and to reduce the stability of the haemoglobin molecule. Homozygotes have only a mild anaemia.

More than a hundred mutations altering single amino acids in globins are known. Most are rare, sometimes restricted to single families. The majority show changes in the stability of the haemoglobin molecule (Figure 1.18). Sometimes the effects are quite severe; the protein precipitates and red cells are trapped in the circulation. Here, as in sickle-cell disease, they are broken down, giving anaemia and – often – spleen damage.

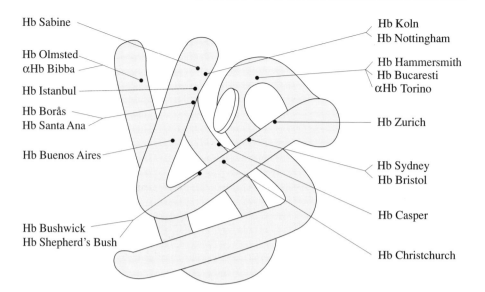

Figure 1.18 Some amino acid substitution sites leading to unstable haemoglobin.

Other single amino acid substitutions have more subtle effects. Some alter the molecule's ability to bind oxygen. Homozygotes for Hb Kansas, for example (a substitution of threonine for asparagine at position 102 in the β chain – not shown in Figure 1.18) often look slightly blue because much of their blood is in the venous (deoxygenated) form. Those with the illness find it difficult to take strenuous exercise.

▷ Why should there be such great differences in the incidence of amino acid substitutions at different sites along the amino acid chain?

▶ Any change in an amino acid which drastically interferes with the molecule's ability to function will not allow its carriers to survive, and will not remain in the population for long. Mutations in the haem-binding part of the molecule are of this kind, and changes in the amino acids in this segment of the haemoglobin chain are almost never found.

The **thalassaemias** are a more complicated group of inherited errors of haemoglobin and can be more serious. They are named from their prevalence around the Mediterranean (from the Greek, *thalassa*, which means sea). All share certain symptoms with sickle-cell, but are genetically quite dissimilar. Instead of a simple substitution of one amino acid for another, one or other of the haemoglobin chains is partly or completely missing (Figure 1.19). They are defects in protein *synthesis* rather than in protein *structure*.

If one of the two chains is not made there will be an excessive number of unpaired copies of the other. Homozygotes for a severe form of β-thalassaemia, for example, cannot make β chains. They will hence have many single α chains which cannot by themselves make a haemoglobin molecule. The chains precipitate as soon as they are produced and interfere with the formation of red blood cells. The few red cells that *are* made are liable to be destroyed.

In α-thalassaemia there is an absence (or a shortage) of α chains. However, the consequences may be less severe than in β-thalassaemia, as the unpaired β chains which remain are more able to lock together to form a reasonably functional molecule made of four β chains – HbH, as it is known. Red cells containing HbH survive less well than their normal counterparts. HbH is less effective in carrying oxygen than is the normal form so the disease may still be a severe one.

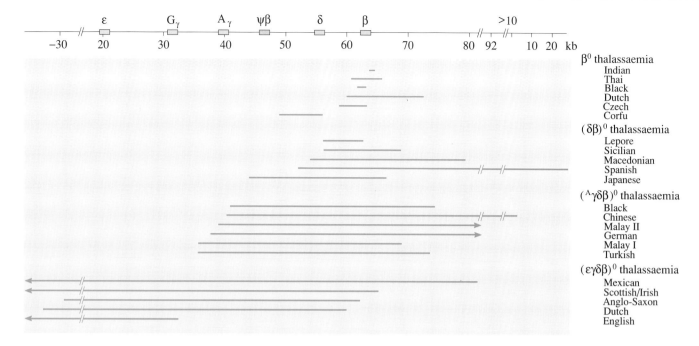

Figure 1.19 Some of the deletions producing β-thalassaemias.

There are more than a hundred different known deletions of the whole or part of particular globin chains. (Figure 1.19 shows just some of the deletions of the β chain.) At the DNA level there are thousands of substitutions in the regions around the globin genes. Some of these may themselves have subtle effects on the phenotype. For example, sickle-cell disease seems to have more severe effects in some regions than in others, perhaps because changes in the DNA closely linked to it are modifying the expression of the gene itself.

The complexity of the relationship between genotype and phenotype is emphasized when it is remembered that hereditary anaemia, which once seemed a simple and unitary disease, can itself have multiple genetic causes. The two thalassaemias share symptoms with sickle-cell disease, but have intricacies of their own. A homozygous β-thalassaemic individual is very anaemic. He or she has great difficulty in making red blood cells, as the excess α chains interfere with the bone marrow cells involved. The body compensates by increasing the activity of these cells, and the arms, legs, skull and face may be greatly deformed as a result. Sometimes the bones even break. The spleen works overtime to clear the abnormal red blood cells, and so becomes enlarged.

The overtaxed cells of the bone marrow demand iron to make haemoglobin. Often, an excess is absorbed from the diet and laid down in awkward places, such as the heart (sometimes causing death from heart failure). Pleiotropy is just as much a feature of the thalassaemias as it is of sickle-cell disease.

1.5.7 The effects of a gene may depend on its environment

Issues of 'nature' versus 'nurture' have been the focus of much social and even philosophical and religious debate, especially in recent years. To what extent is a particular characteristic – disease, musicality, or tendency to commit a crime, for example – innate, and how much can be modified by the environment? Yet geneticists should know that this superficially intriguing issue is often impossible to resolve. Indeed, it often has no meaning: the only answer to the nature–nurture argument is that there is no sensible question.

Many of the symptoms of sickle-cell disease arise through a failure to transport oxygen properly. As a result, patients are at greater risk in conditions where oxygen is in short supply; that is, in one environment rather than another. Heterozygotes usually have fewer problems. However, there are circumstances in which even they may be in danger of showing some symptoms of the disease. Although commercial aircraft pose few dangers for carriers, those with a single copy of the sickle-cell allele (S) are advised not to travel in military planes which may be poorly pressurized. Some have had problems after receiving an anaesthetic or a tourniquet applied to stop bleeding from a wound. Even for A S heterozygotes (most of whom know nothing of their genotype) there may be tissue damage under these circumstances. Once again, both innate and external factors are involved: the damage arises not only because of the change in the environment brought about by flying or by anaesthetics, nor solely because of their genotype. It happens when both are brought together – nature and nurture in combination.

▷ Women are more at risk from the effects of the condition than are men. Why might this be?

▶ Pregnant women with sickle-cell disease suffer more because of the demands of their fetus for oxygen. Many pregnancies are lost.

The whole of the medical treatment of inborn disease, of course, depends on the fact that a phenotype can be transformed by changing the environment in which it is placed even though the genotype of the patient remains unaltered. Unfortunately, therapy for the *haemoglobinopathies* (the diseases resulting from damage to haemoglobin) is not very successful. Some of their most virulent effects can be alleviated. Many sickle-cell patients are in danger of infection which can be held at bay with antibiotics. The anaemia itself may be helped by giving a diet rich in a particular vitamin, *folic acid* (which is important in red blood cell formation). Repeated blood transfusions may help, too. Dozens of drugs have been tried in the hope of reducing the tendency of red cells to sickle. All have failed: this aspect of the phenotype has, so far, proved resistant to treatment. When an effective drug is found, one of the most direct phenotypic effects of the sickle-cell allele – the deformation of red blood cells – will be altered by a planned change in the environment. Careful nurturing will, with luck, ameliorate the effects of a defective nature.

1.5.8 *Dominance and recessivity are not as simple as they seem*

To the early geneticists, dominance and recessivity were straightforward; they were intrinsic properties of particular alleles. We now know that they are attributes of phenotypes, not genotypes, and are more subtle than once appeared. For example, sometimes there is no dominance and a heterozygote manifests the effects of both the alleles it possesses. Often, one allele may be dominant for some of its attributes and recessive for others.

For sickle-cell, it has long been known that some of the attributes of the mutated gene are recessive and others dominant. Heterozygotes show none of the painful symptoms of those with the disease and the S allele is *recessive* in these respects: A S individuals are normal for most characteristics. Some heterozygotes, though, have a slight haematuria; that is, haemoglobin appears in the urine. This is also one of the symptoms of those with the disease. In this respect, the A S genotype has a phenotype similar to S S; that is, S is *dominant* in its effects on haematuria.

To complicate the issue further, heterozygotes may begin to show some symptoms of the disease if they are placed in conditions of low oxgyen concentration.

▷ Why might the existence of dominant attributes of the phenotype in diseases whose effects are in general recessive to those of the normal allele be of practical use to geneticists?

▶ It might help in genetic counselling as it can allow heterozygotes to be detected.

With the advent of electrophoresis (Book 1, Chapter 3), and the more sophisticated methods of molecular biology which succeeded it, our view of dominance relations in sickle-cell and other inherited diseases has changed. Such technical developments have been of great importance for genetic counselling. Now heterozygotes may be detected by looking at the gene products themselves. Electrophoresis can be used to detect the slight differences in the charge of a protein molecule which may arise if one amino acid is substituted for another. By applying a powerful electric current to a blood sample placed on a gel material which acts as a filter, haemoglobins differing in charge are drawn to differing extents into the gel. Figure 1.20 shows the patterns from a person with sickle-cell disease (genotype SS), a heterozygote (AS) and a normal individual (AA).

Figure 1.20 shows that the heterozygote has one copy of each protein product: a band for HbA and one for HbS. At this level of investigation, closer to the gene itself, there is no dominance of one allele over the other. Both gene products can be seen on the gel. This is **codominance**.

At the DNA level, codominance is the rule. Much genetic counselling – identifying heterozygotes in the hope of advising them on the chances of having an affected child – depends on the ability to penetrate the phenotypic expression of a gene to detect genetic damage directly. Carriers can then be identified by looking at the DNA sequence – where dominance and recessivity vanish – rather than at their phenotype (which is usually indistinguishable from that of normal individuals).

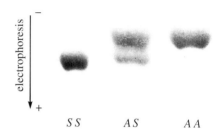

Figure 1.20 Result of electrophoresis of haemoglobins from individuals with genotypes SS, AS and AA (see text).

This helps, too, in cases in which particular genes are switched off in the embryo (as is, for example, β-globin) and produce no protein. Looking at DNA damage can identify the genotype of the fetus – even though it is impossible to identify the gene product itself – and the prospective parents can be informed. In cases in which the location of the structural gene responsible for an inherited disease is not known it may be possible to track down a DNA variant that travels down the generations in consort with the disease, and to use this to infer the genotype of a developing embryo or of someone who is at risk of carrying a single copy of the damaged allele.

1.5.9   The effects of one gene may be modified by those of another

Any organism is the result of a complex web of interacting genetic and developmental pathways. A change at one locus often has effects on others. The haemoglobins are no exception. A mutation in one of the genes may be ameliorated by a compensating alteration in another.

▷ In Cyprus and other places both α- and β-thalassaemia are found in the same population. Most of the symptoms of the two thalassaemias are due to the presence of a pool of α- or β-globin chains lacking counterparts because the matching gene has been damaged. How severe do you expect the symptoms to be in a person in which *both* chains are affected?

▶ If *both* genes are damaged, neither works with full efficiency and there is no great excess of single chains. People with both deletions – joint α- and β-thalassaemia – hence show less severe symptoms than do those with only one. This is an unusual case in which two genetic defects have a joint effect less severe than does either alone.

Other interactions between genes can also be important. During development from egg to adult there is a shift in the activity of the globin genes. Fetal haemoglobin (HbF) is usually found only in tiny amounts in adults; the gene involved is not required as normal adult haemoglobin is best adapted to oxygen transport between lungs and tissues. However, if there is a fault in the synthesis of either of the chains that make up adult haemoglobin, the balance can change.

Children with severe thalassaemia often survive better than might be expected given that they are almost unable to make normal haemoglobin. Fetal haemoglobin has come to the rescue. It is produced for far longer into adult life than is normally the case. Often, the effect is due to the increased survival of those rare cells in the bone marrow which contain HbF, as their competitors die because of their failure to complete haemoglobin synthesis. Once again, the phenotype is the result of a complicated interaction between distinct gene loci.

1.6 Envoi – the moral of the sickled cell

The haemoglobinopathies provide a model of just what the new genetics can do. Surely, the optimist might argue, all this must mean that a cure, or at least a treatment, for haemoglobin disorders is near at hand. In fact, the opposite is true. The scientific advances made in studying the molecular genetics of haemoglobin have illuminated most areas of genetics, from gene structure to gene regulation, mutation and evolution. With the important exception of the development of tests for prenatal diagnosis, though, they have made little difference to the treatment of its disorders (let alone any cure).

There is hope that an understanding of the biochemistry of the diseases may help. For sickle-cell, the most widespread of the haemoglobinopathies, there are several potential strategies. Perhaps genetic engineering will make it possible to design short peptides that can be synthesized in the red cells and might interfere with the formation of the sickle-cell haemoglobin fibres. Another approach may be the use of agents that alter the permeability of the red-cell membrane, thereby allowing more water to enter and so reducing the chance of fibre formation. It is well known that cyanates increase the oxygen-binding capacity of haemoglobin and, at least *in vitro*, reduce the number of fibres formed and hence protect red cells against sickling. Cyanates have been tried on patients, but unfortunately have dangerous side-effects. Among more drastic options there is the hope of persuading fetal haemoglobin to be synthesized during adult life, giving a new source of haemoglobin (as happens naturally in some thalassaemia patients). All this lies in the future – as does gene therapy for most other inherited diseases.

For the present, screening for heterozygotes or for affected fetuses and aborting them is the most effective way of reducing the incidence of abnormal haemoglobins. Unfortunately, the screening programme for sickle-cell carriers which began in the United States two decades ago is now seen to have been a disaster, albeit one embarked upon with the best of motives. The millions of Americans who carry a single copy of the sickle-cell gene are in most circumstances, perfectly healthy. Many have no idea of their genetic status.

Soon after a test for carriers became available there was pressure in some States for the test to be made compulsory (the reason why was never very clear). Those found to be heterozygotes were discriminated against when it came to obtaining health insurance (which, in the absence of a national health system, is essential in the USA) or even in employment. Carriers often suffered from feelings of guilt, of misery and of ill-health. Relationships broke down as those bearing a single copy of the gene were perceived to be undesirable marriage partners. Astonishingly enough, there were even cases in which children with the disease suffered emotionally as well as physically, because the illness was thought to be contagious and those affected with it best avoided.

Counselling programmes elsewhere in the world have been more successful although, sometimes, they depend on quite considerable social pressure. In Cyprus and in Italy a combination of prenatal diagnosis and of advice to intending marriage partners about their carrier status has reduced the number of thalassaemic children born. Elsewhere in the world, though, such programmes have not yet had a chance to demonstrate their effectiveness. 'Effectiveness', too, is often not well defined. There are issues of personal choice and of ethical belief which must be balanced against potential economic or medical benefits. As in much of medicine, education of the public (often involving, in these societies, the cooperation of the church) is crucial to the success of any counselling programme.

Not all the news about sickle-cell disease is bad: but most of the good news for those suffering from the condition comes from conventional medicine. At least in the affluent West, the health and survival prospects of sickle-cell homozygotes has improved greatly over the past 20 years. In 1973, the median survival was only 14.3 years, and one SS baby in five died before its second birthday. The chance of living to be more than 30 was only one in six. In 1993, though, 85% of children with the disease lived to be more than 20; and most patients survived until their 40th birthday. Many of the earlier deaths were due to chest infections which can now be controlled with antibiotics, or to the failure of kidneys (which may be replaced with a transplant). Geneticists have had a disappointingly small part to play in improving the lives of those with the disease.

All this is a microcosm of the problems that can arise when public perceptions do not accord with scientific reality. However, although studies of haemoglobin may not have produced the 'quick fix' which medicine is always searching for, they have given a unique insight into the structure of genes and the nature of mutation. This insight may, ironically enough, be of more help in treating other genetic diseases rather than those of the haemoglobins themselves.

Objectives for Chapter 1

After completing this chapter you should be able to:

1.1   Define and use, or recognize definitions and applications of, each of the terms printed in **bold** in the text.

1.2   Discuss the history of genetics and what genetics can and cannot say about inheritance, recalling the work of Mendel and his laws of segregation and of independent assortment.

1.3 Describe the evidence that places genes on chromosomes and how chromosomal abnormalities are the proof of the chromosome theory.

1.4 Interpret simple human pedigrees and identify dominant, recessive and sex-linked inheritance in humans, using the information to estimate genetic risks.

1.5 Discuss the human globin genes: their structure and normal function; various kinds of errors from point mutations to deletions; their patterns of activity from early embryo to adult; the existence of genetic errors – mutations – as one generation succeeds the last; the manifold effects of even a single base change in the DNA; the interaction of nature and nurture; the complexity of the notions of dominance and recessivity; and the interaction of different gene loci in determining the phenotype.

1.6 Understand the limitations of genetics in treating (let alone curing) the majority of inherited diseases.

Questions for Chapter 1

Question 1.1   (Objectives 1.1 and 1.2)

After completing the first two generations of his experimental crosses with peas, to check his results, Mendel did another experiment. He self-fertilized the F_2 yellow-seeded plants emerging from his first experiment (p.5). Of 519 individuals, 166 bore only yellow seeds in the next generation, the F_3. The remainder produced both yellow- and green-seeded offspring. Adding together all the F_3 plants gave the usual 3 : 1 ratio of yellow to green; but some of the F_2 yellows were, unlike their parents, pure-breeding. An apparently 3 : 1 ratio in the F_2 could more accurately be described as a 1 : 2 : 1 ratio. What are the genotypes of the F_2 plants in this experiment?

Question 1.2   (Objective 1.4)

There are a few X-linked *dominant* conditions in humans, although they are rare. Figure 1.21 gives a (slightly simplified) pedigree of phosphate diabetes, a condition that involves disturbances of both phosphate and sugar metabolism.

(a) What is the evidence that the condition is indeed an X-linked dominant one?

(b) What are the probable genotypes of individuals II-2, III-4 and III-5?

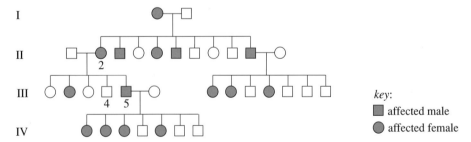

Figure 1.21 Pedigree showing four generations of phosphate diabetes inheritance.

key:
■ affected male
● affected female

Question 1.3   (Objective 1.6)

In the distant land of Albany, one child in 3 600 is born with a recessive autosomal condition that removes its skin pigment and means that it needs expensive sunblock cream before it can safely play outside. In the interests of reducing future expenditure, Albany's dictator decrees that such children should not be allowed to reproduce. How effective will his policy be in achieving his aims? You should disregard, for the time being, the ethical issues involved.

The map of the genes

This chapter is concerned with how genes can be localized: the basic task of genetics. Finding a gene is the first step in identifying what that gene does. This is important, especially in cases where alterations – mutations – of a gene will lead to disease. In particular, finding the genes responsible for inherited diseases may allow their function to be understood and might eventually lead to improved treatment, or the possibility of prevention. Many of the techniques that you will learn about in this chapter will be met again in Chapter 8 which describes the cloning of the cystic fibrosis gene.

It will be particularly useful to watch Video sequence 4 both before and after reading this chapter.

2.1 Chromosomes

2.1.1 Chromosome structure

Using a light microscope, chromosomes can only be seen clearly during cell division, when they become compacted up to 10 000-fold. If the chromosomal DNA of a single human cell were to be stretched out it would be over 2 metres long. In the cell, several levels of condensation occur. For most of the time the DNA exists as thin threads of **chromatin**. This consists of a DNA strand wrapped around proteins (histones) to form **nucleosomes** (Figure 2.1). Under the electron microscope these look like beads on a string, with stretches of linker DNA between each nucleosome.

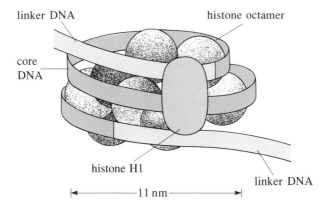

Figure 2.1 Nucleosome structure. A 146 base pair length of DNA, the core DNA, is wrapped around eight histone protein molecules (histone octamer) to form each core particle. DNA stretching between the nucleosomes is called linker DNA. This arrangement is often referred to as 'beads on a string'.

These 'strings of beads' are wound into helical arrays or 'solenoids' (Book 1, Figure 5.18) that reduce the length of the DNA to one-fortieth of the original. To organize these long fibres during cell division would be difficult, and in dividing cells the chromatin is further condensed to around one ten-thousandth of its original length. This is achieved by the fibres being attached in loops to a protein scaffold in association with histone proteins. Figure 2.2 shows how the mass of DNA falls away from the protein scaffold when the histones are removed.

It was only in the mid-1950s that the number of human chromosomes in a diploid cell was shown to be 46. Since then there have been enormous technical advances. The chromosome complement of an individual is known as the **karyotype** (Figure 2.3). Recent developments in staining (treating the chromosomes with dyes such as Giemsa so that they can be visualized) now allow cytogeneticists

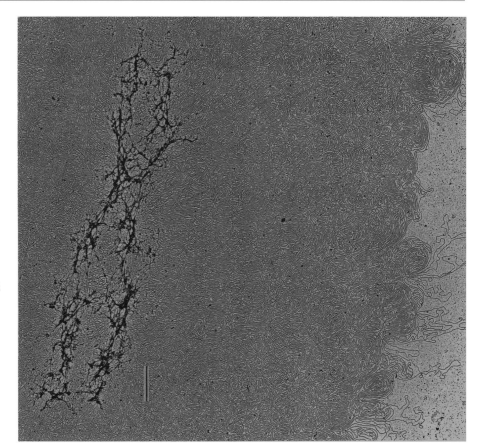

Figure 2.2 A human chromosome with the histone proteins removed. The dark-staining central scaffold outlines the shape of the two chromatids which are joined at the centromere. After removal of the histones the DNA loops out from the scaffold in a tortuous maze. In theory, it is possible to trace a loop out from the scaffold and back again. (Scale bar = 1 μm.)

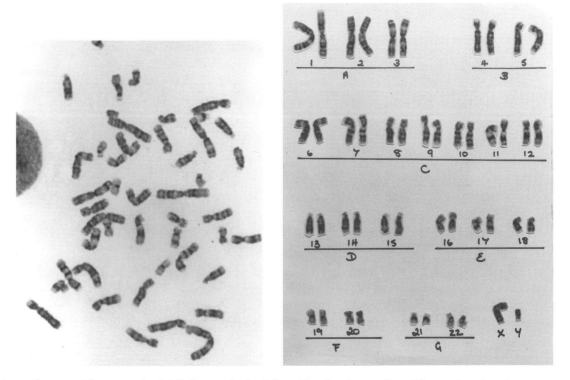

Figure 2.3 The karyotype of a normal human male. (a) Cells are obtained from blood and are cultured (grown) so that they divide. Following treatment with a chemical that traps them in metaphase, they are dropped onto a microscope slide and fixed there. In some cells the dividing chromosomes spread nicely so that they can be identified. Staining helps each chromosome to be recognized as they all have characteristic patterns of light and dark bands. (b) Chromosomes from the photograph in (a) have been cut out and arranged in approximate size order: 1–22 and then the X and Y chromosomes. (Females have two X chromosomes.)

(people who study chromosome structure) to see light and dark bands on each chromosome, and so identify particular chromosomes on the basis of their characteristic banding patterns.

The staining pattern of particular regions reflects nucleotide composition; relatively (A+T)-rich regions stain darkly (Giemsa-dark bands) and (G+C)-rich regions stain lightly (Giemsa-light bands). Giemsa-dark bands contain relatively few genes. In fact, about three-quarters of those mapped are found in Giemsa-light bands.

2.1.2   Abnormal karyotypes

Karyotype analysis enables alterations in the number or size of chromosomes to be detected. There may be an extra copy of one whole chromosome, or one copy may be missing. An alteration in chromosome complement is termed **aneuploidy**. Aneuploidy can often be correlated with particular syndromes. For example, Down's syndrome is the result of an extra chromosome 21 (**trisomy 21**). Down's syndrome individuals have varying degrees of mental retardation, are shorter than average and have a number of characteristic facial features. In addition, leukaemia is more common in these individuals than in the general population. Other trisomies exist, but only those of the smaller autosomes ever survive to birth. Trisomy 18 (Edwards syndrome) and trisomy 13 (Patau syndrome) both result in many abnormalities and lead to early death. (Interestingly, trisomies of the sex chromosomes are tolerated fairly well.) Chromosomes 13, 18 and 21 are found to have quite a high proportion of Giemsa-dark bands.

▷   What feature of chromosomes 13, 18 and 21 may explain the viability of trisomies for these chromosomes?

▶   They are among the smaller chromosomes (21 is the smallest). Also, the observed preponderance of Giemsa-dark bands suggests that these chromosomes carry the fewest genes, so causing least disruption when present in excess.

A lack of one copy of a chromosome (**monosomy**) is less readily tolerated than is trisomy. This has been well demonstrated in *Drosophila* where gene content can be manipulated experimentally. Flies with three copies of one-tenth of their total DNA are viable. However, losing one copy of just one-thirtieth of the total DNA is lethal. An example of monosomy in humans which does not cause death is Turner's syndrome in which there is only one X chromosome and no Y, giving an 'XO' karyotype. Children with this syndrome are female and survive; this is not surprising since all males manage with just a single X chromosome. Genes that would normally be expressed from either the second X chromosome (in females) or the Y chromosome (in males) must be responsible for preventing the clinical features seen in Turner females. They are short and have poorly developed sex organs. Some have abnormalities of the aorta and webbing of the neck.

Chromosomal abnormalities (which are discussed further in Chapter 4) have been useful in determining where on the chromosomes various genes are found. For example, genes that when present in three copies result in the features of Down's syndrome must be present on chromosome 21. Additional symptoms of this disease include relatively rapid ageing and, sometimes, symptoms of mental disturbance similar to those found in the form of pre-senile dementia known as Alzheimer's disease. This was one of the clues that led to the discovery that the gene for a form of Alzheimer's disease that runs in families is indeed on chromosome 21.

Further examples of the ways in which chromosome abnormalities can be used to locate genes will be described in Section 2.3.3.

2.2   In search of landmarks

This section describes the ways of assigning or **mapping** a gene to a particular chromosome and then shows how its position can be tracked to smaller and smaller regions of the chromosome until the gene itself can be pinpointed.

There are two ways in which it is possible to localize or map genes and other DNA sequences:

1   To make use of the fact that recombination occurs during meiosis. The extent of recombination, determined by analysis of pedigrees, is used as a measure of the **genetic distance** between genes and can be used to produce a **genetic map**.

2   To deal with the DNA sequence directly and to make a **physical map**. The ultimate physical map of an organism is its complete DNA sequence: the order of bases in the whole length of its genetic material.

These two complementary approaches to gene mapping will be discussed in this and the following chapter.

Figure 2.4   Genetic linkage. Recombination occurs at meiosis when the two homologous chromosomes are lined up alongside each other. At this stage in the cell cycle the DNA has been replicated so that each chromosome consists of two strands or *sister chromatids*. Recombination occurs between only one chromatid from each of the homologous chromosomes, due to physical constraints. When an individual is heterozygous at two loci, the combination of alleles in the gametes shows whether recombination has occurred between the loci. (a) When two loci are very close together it is unlikely that recombination will occur between them and so the alleles are passed to the gametes in the same combination as in the parent. (b) If the two loci are far enough apart so that one recombination event always occurs between them, the alleles are passed on to the offspring in all possible combinations (Mendel's law of segregation). In the case shown, the loci are on the same chromosome but from the patterns of recombination it cannot be distinguished whether the loci are far apart on the same chromosome or on different chromosomes. If loci are sufficiently far apart for a recombination event to occur between them sometimes *but not always*, then the proportion of offspring that are recombinants gives an indication of the distance between the loci.

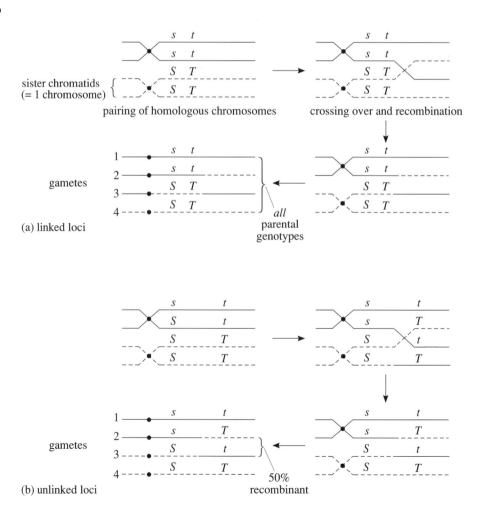

2.2.1 Genetic maps

The principles of **linkage mapping** have been used to map genes in experimental organisms such as *Drosophila* and mice. In humans the process is not as easy due to the small family sizes and the inability to set up particular matings. However, despite these difficulties, linkage mapping has been very successful in humans. The rationale behind the procedure is described below.

As they have two copies of each autosome, diploid organisms have two copies of every autosomal DNA sequence. Individuals are heterozygous if they have different DNA sequences (alleles) at one position (or locus) on each of the chromosomes of an homologous pair. If the individual is also heterozygous at another locus on the same chromosome then it is possible to determine whether recombination has occurred between the two by looking at the combinations of alleles in the offspring.

If two loci are close together on the chromosome then there is less distance within which recombination may occur. The alleles will hence tend to be inherited in the same combinations as they were on the parental chromosomes. Such loci are **linked** (Figure 2.4a). If the loci are far apart or on separate chromosomes, then they will recombine freely and the four alleles will be inherited in all possible combinations. Such loci are **unlinked** (Figure 2.4b). The extent of linkage of two loci is thus an indication of the genetic distance between them.

▷ Draw a diagram similar to Figure 2.4 to illustrate what gametes are produced when an individual is heterozygous at only one of the loci when they are (a) linked and (b) unlinked.

▶ A parental genotype heterozygous for one of the loci is ST/St. Figure 2.5a shows the gamete genotypes produced when the loci are linked and Figure 2.5b the results for unlinked loci.

▷ In the case above, can it be determined whether recombination has occurred?

▶ No; as there are only two classes of gametes, ST and St, both of which are the same as the parental combination and could have arisen either from recombination or from an absence of recombination.

▷ The maximum amount of recombination seen in gametes (that is, the number of recombinants divided by the number of non-recombinants plus recombinants) is 50%. Why is this?

▶ Look at Figure 2.4b again. The two loci are so far apart that a recombination event always occurs between them. However, as the recombination only occurs between one set of non-sister chromatids, half of the gametes still have the alleles in the parental combination. Thus the proportion of recombinants can only be 50%. When two loci are on different chromosomes the same is true as the chromosomes are equally likely to assort in all possible combinations.

Note that in these examples we are only looking at the products of meiosis of one of the parents. When examining alleles in the offspring it is vital to determine which allele combination has been inherited from the affected parent. This is much easier if the second parent is homozygous or has different alleles from the first parent at both of the loci being examined.

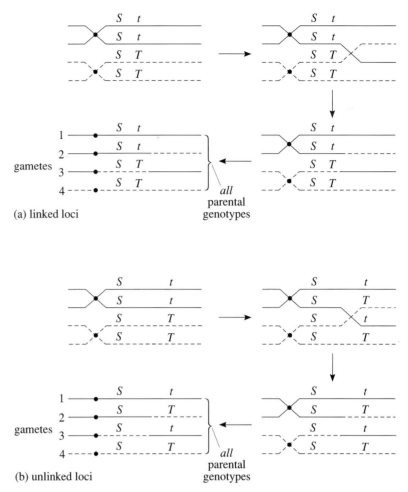

(a) linked loci

gametes
1 — S t
2 — S t
3 — S T
4 — S T

all parental genotypes

(b) unlinked loci

gametes
1 — S t
2 — S T
3 — S t
4 — S T

all parental genotypes

Figure 2.5 See text.

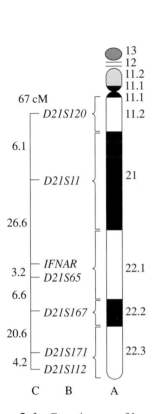

Figure 2.6 Genetic map of human chromosome 21. The diagram shows a representation of the banding pattern of chromosome 21 (column A). Each band is numbered so that geneticists can exchange information about particular chromosome regions. Regions on the short arm are given the prefix 'p' and those on the long arm 'q'. (The centromere is at position 11.1.) Names of some loci are shown in column B. Locus *D21S167* (21 for chromosome 21, S for single copy) is in chromosome band 21q22.2. In column C are the distances between loci in centimorgans. It is interesting to note that many copies of the genes for 18S and 28S rRNA (of the small and large ribosomal subunits repectively) are found in band 21p12; multiple copies of these genes are also present on the short arms of chromosomes 13, 14 and 15. The region 21p13 is called a 'satellite' (see Chapter 5) and contains repetitive sequences.

Linkage analysis, the study of the patterns of inheritance of loci through generations of a family, is used to build up a genetic map depicting the relative distances between different loci based on the frequency at which they recombine with each other.

Genetic distance is measured in **centimorgans (cM)**. If a recombination event occurs between two loci in one out of every hundred meioses (i.e. 1% recombinants) they are said to be one centimorgan apart. In humans, one centimorgan corresponds, very approximately, to a physical distance of 1 million base pairs. A genetic map of chromosome 21 is shown in Figure 2.6.

2.2.2 An excursion into polymorphism

In each diploid cell, there are two copies of every gene, one on each chromosome. In some cases both copies are identical in every individual in a population. The gene or locus is then said to be **monomorphic**. Sometimes, though, there is more than one form of a gene. Such a locus has more than one allele and is polymorphic (Chapter 1). Think of the gene determining seed colour from Mendel's experiments. The seeds were of two colours (green or yellow) and the gene has two alleles. In humans, the ABO blood group locus has three alleles, and an individual possesses any two of them. In many cases, one allele will be dominant to the other, so only one will show its effect in the phenotype (thus, for example, the seeds of a pea plant are either green or yellow, but not a mixture of both).

▷ The genes of the ABO blood group system direct the addition of carbohydrate chains to antigens on the surface of the red blood cells. The alleles are codominant; in other words, neither is dominant to any other. However, the *O* allele does not encode the addition of any carbohydrate and is effectively a **null allele** (one without an expressed phenotype of its own). Write down all the possible combinations of the three alleles and infer the phenotype of each combination.

▶ The table below lists the blood group genotypes and the corresponding phenotypes.

Genotype	Phenotype (blood group)
O O	O
A A	A
A O	A (the *O* allele has no effect)
B B	B
B O	B
A B	AB

For simple Mendelian analysis of single genes, only the locus under investigation need be polymorphic: indeed, without polymorphism classical genetics would be impossible.

As you have seen, for linkage mapping, a parent must be heterozygous at two or more loci to determine whether recombination has occurred. The most important prerequisite for such analysis, therefore, is to identify polymorphic loci. In addition, it is helpful to use loci at which the alleles are present in the population in more or less equal proportions so that many individuals will be heterozygous. The **heterozygosity** of the locus is the proportion of individuals in a population who are heterozygous. Any mating in which the parent is not heterozygous at two or more loci is not **informative** for linkage mapping. In humans, where matings are not controlled, it is important to find as many polymorphic loci as possible to make as many of the families as possible informative.

If the frequencies of two alleles are known (as a result, for example, of surveying the blood groups of a population), the expected number of heterozygotes can be calculated using the Hardy–Weinberg formula (see Chapter 1).

▷ (a) What will be the heterozygosity if the frequency (p) of allele *a* is 0.3?

(b) What will be the heterozygosity if the frequency of homozygotes for the *a* allele (*a a*) is 0.09?

▶ The answer is the same in both cases – the question is simply phrased differently:

(a) The frequency of heterozygotes is $2pq$. In this case $p = 0.3$, $q = 1 - p = 0.7$; so $2pq = 0.42$.

(b) The value of p is $\sqrt{0.09}$, since p^2 gives the number of homozygotes *a a*. Therefore $p = 0.3$ and the number of heterozygotes can be calculated as above.

2.2.3 An example of genetic linkage : nail–patella syndrome and ABO blood groups

Figure 2.7a illustrates how a disease locus can show linkage to another polymorphic locus in a family. It shows the pedigree of a family having some members affected by the autosomal dominant condition called **nail–patella syndrome**. This disease has three major clinical features; abnormal growth of the nails, absence of the patella (knee bone) and the presence of characteristic bone structures in the pelvis. Three generations are represented by the numerals I, II and III at the left-hand side. As is the usual convention, males are represented by squares, females by circles, and affected individuals are shown shaded. Each family member can be described by a number. For example, the female II-8 is affected. The letters to the bottom right of each symbol are the alleles of the ABO blood group locus for each individual.

Note one striking feature of this pedigree: every affected individual has inherited the B allele from individual I-2 along with the disease allele, i.e. the two characters **co-segregate** in the pedigree.

The disease is not always inherited with the B allele: in other families nail–patella syndrome may be inherited with either the A or the O allele. All this suggests strongly that the nail–patella locus is close to that for the ABO blood groups.

▷ Look at the pedigree in Figure 2.7b. Which blood group allele accompanies the gene for nail–patella syndrome? What has happened during gamete formation in individual II-1? (*Hint*: look at offspring III-2).

Figure 2.7 Pedigrees showing the close genetic linkage between the nail–patella syndrome locus and the ABO blood group locus.

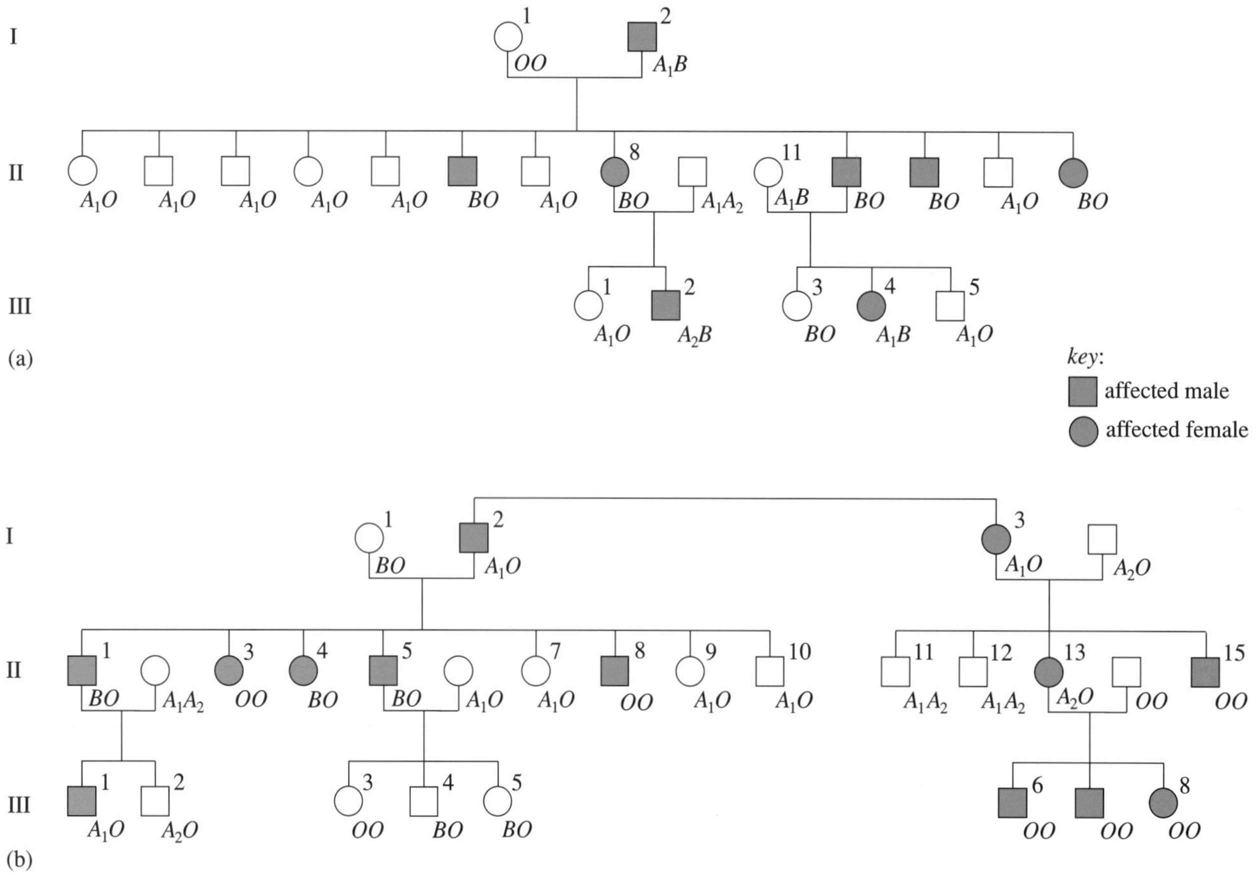

▶ The gene is, in most individuals, inherited along with the *O* allele. There has been a recombination event between the ABO blood group locus and the disease locus during gamete formation in II-1 so that III-2 has the *O* allele but not the disease. The same thing has occurred in II-5 so that III-3 has both *O* alleles from the parents but does not have the disease.

In a larger study involving several families the percentage recombination (the number of recombinant individuals divided by the total number of offspring) between the ABO blood group locus and the nail–patella syndrome gene was 10%. This is the proportion of meioses showing recombination (one recombination for every ten meioses).

▷ Based on the convention that one centimorgan represents one recombination event per one hundred meioses, how far apart are the loci?

▶ The two loci are 10 cM apart.

The above relationship does not work for longer distances, because of the probability of *two* recombination events between the loci. For short distances, though, it is reasonably accurate to regard centimorgans and per cent recombination as equivalent.

The requirement of heterozygosity for linkage mapping makes it important to have available as many polymorphic sequences as possible. Some polymorphic loci – eye colour, for example – can be seen directly, while others can be detected at the biochemical level (as in the case of blood groups). However, polymorphism at the DNA level was impossible to detect until recently. The remainder of Section 2.2 describes how polymorphism can be identified in the laboratory using new techniques, and used in family studies to generate genetic maps.

2.2.4   Restriction fragment length polymorphisms (RFLPs)

A major breakthrough in the field of molecular biology was the discovery in the 1970s of **restriction enzymes**. These enzymes are isolated from bacteria where they destroy the DNA of invading foreign organisms. They cleave DNA at a precise site within a specific recognition sequence which is different for each restriction enzyme, as described in Box 2.1.

Box 2.1   *Restriction enzymes*

Restriction enzymes are endonucleases that cleave both strands of the DNA at a specific sequence. They are the 'scissors' of the cell. The cleavage sites are characterized by the nature of the sequence. This always has two-fold rotational symmetry; that is, if it is rotated through 180° (turned upside-down) it reads the same as it did before the rotation. The sequence is therefore a palindrome.

For example, the recognition sequence for the enzyme *Bam*HI is:

5′ G↓G A T C C 3′

3′ C C T A G↑G 5′

The position where the DNA strand is cut is marked with an arrow. The notation 5′ (5 prime) and 3′ (3 prime) refer to the direction, or polarity, of the DNA strand.

The two strands run in opposite directions to each other.

The names of restriction enzymes are derived from the bacteria they come from. *Bam*HI, for example, is isolated from *Bacillus amyloliquefaciens* H.

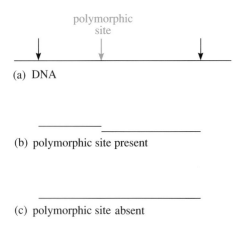

(a) DNA

(b) polymorphic site present

(c) polymorphic site absent

Figure 2.8 Restriction fragment length polymorphism.
(a) Arrows indicate the sites of cleavage for a restriction enzyme in the DNA.
(b) If the central polymorphic site is present two fragments will be produced.
(c) If, due to a change in the DNA sequence, the central site is abolished, a single fragment results.

Restriction enzymes have been used to reveal extraordinary amounts of polymorphism in human DNA and are also used in cloning DNA, which is part of the process of *physical* mapping (see Section 2.3).

Since the enzymes will only cleave the DNA within a precise sequence, any base-pair change within the recognition sequence will stop them working: there will be no cleavage at that site. The pattern of digestion products of samples of DNA from different individuals will therefore differ if there is a DNA polymorphism that creates or destroys an enzyme recognition site. The pieces of cut DNA are separated by electrophoresis through an agarose gel. This separates them by size (see Book 1, Box 5.2). A change in restriction enzyme sites creates longer or shorter fragments. The term **restriction fragment length polymorphism** or **RFLP** is used to describe such variation (Figure 2.8). The detection of RFLPs is described in Box 2.2. and Figure 2.9 (see facing page).

▷ In the example given in Figure 2.8 a probe in this region will detect just one polymorphic site when the DNA is digested with that particular enzyme. Although the probe itself may be short it will detect the whole fragment that it hybridizes to. In this case the probe is to the right of the restriction site; it will be shorter when the site is cleaved and longer when the site is not cleaved. As an individual has two copies of every sequence, there are three possible genotypes at this locus: homozygous for the presence of the site, homozygous for its absence, or heterozygous (one gene copy has the site and the other does not). How will these three genotypes appear on a gel?

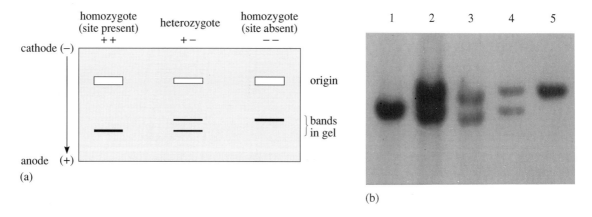

(a)

(b)

Figure 2.10 RFLP analysis by autoradiography. (a) Relative positions of bands for: homozygote for the presence of the polymorphic site, homozygote for its absence, and heterozygote.
(b) Actual autoradiograph of RFLPs from five individuals. Note that bands 2, 3 and 4 all look slightly different as there is more DNA loaded on the gel for some individuals that for others (2 in particular). This makes the bands more intense and the DNA moves at slightly different rates through the gel.

▶ The bands on an agarose gel corresponding to the three different genotypes will appear as in Figure 2.10a.

Figure 2.10b is an actual autoradiograph of RFLPs:

individual 1 is homozygous for presence of site

individuals 2, 3 and 4 are heterozygous

individual 5 is homozygous for absence of site

The existence of RFLPs has revolutionized human gene mapping. They provide a source of polymorphic loci whose inheritance patterns in families make it possible to determine, as for any other locus, genetic distances between them. Some may be inherited independently and may be on different chromosomes. Others, though, may be inherited together – they are closely linked. By looking at many different markers in many families it becomes possible to make a linkage map including both RFLPs and clinically important loci. Ideally, the same

Box 2.2 Detection of RFLPs

A sample of genomic DNA is digested with restriction enzymes and then separated according to size by electrophoresis through an agarose gel. It is then bathed in ethidium bromide which attaches itself to the DNA and fluoresces under ultraviolet light. The resulting smear of many different-size pieces of DNA does not, at this stage, allow any individual RFLP to be detected. Figure 2.9a shows a diagram of such a gel and alongside it is a photo of an actual gel (note the DNA size markers – these are for calibration).

After treating the gel with an alkaline solution to separate the two DNA strands, the DNA is transferred from the gel to a nitrocellulose filter. The transfer of DNA from a gel to a filter is termed a **Southern blot**, after its inventor Ed Southern (Figure 2.9b).

The filter is then carefully peeled off the gel (Figure 2.9c). The DNA is now on the filter and its pattern on the filter is a replica of that on the gel. This single-stranded DNA can be annealed, or **hybridized**, to short lengths of single-stranded DNA with complementary sequences to form a double-stranded molecule. The DNA sequences used are called **probes**. (See Figure 2.9d and e.) The probes are previously labelled with radioactivity or light-emitting chemicals so that the genomic DNA fragments and their associated probes can be visualized by exposing the membrane to X-ray film (**autoradiography** – Figure 2.9f).

Only the cut pieces that the probe hybridizes to will be visible. In this way a subset of the many DNA fragments is selected. Probes may be used to detect a region of DNA where an RFLP is known to exist. A polymorphic DNA segment of this kind can be used in genetic mapping and is termed a **marker**.

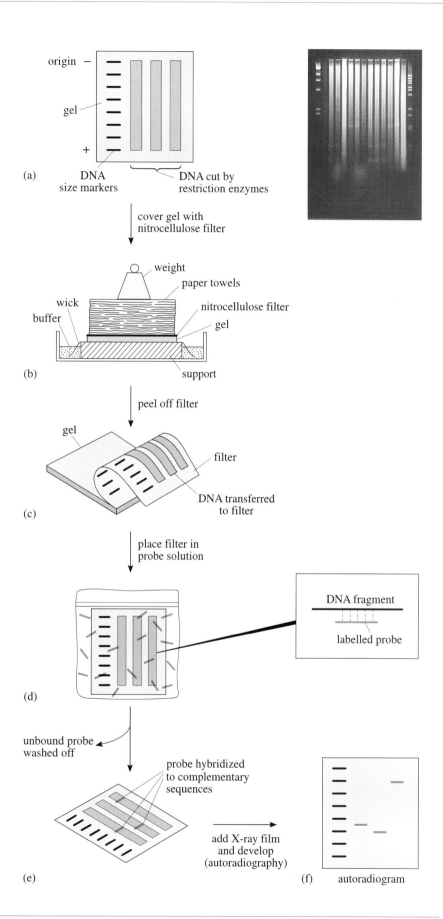

Figure 2.9 Protocol for detection of RFLPs.

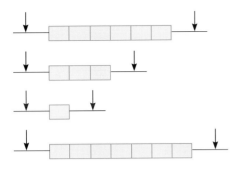

Figure 2.11 Detection of variable number tandem repeats (VNTRs) or *minisatellite* loci (see later). The shaded boxes represent repeated sequences. A restriction enzyme that cuts outside these repeated blocks will produce larger or smaller DNA fragments depending on the number of repeated units – this number varies between individuals.

families should be examined with each probe so that all RFLPs can be ordered with respect to each other. To this end, the Centre d'Etude du Polymorphisme Humain (CEPH) in Paris has collected immortalized cell lines (cells, obtained from an individual, that can be grown indefinitely) from a large number of three-generation (grandparent, parent, children) families. DNA samples from these are widely used for mapping DNA sequences. The information concerning inheritance patterns for many markers is stored on a computer database. The data are readily available and new markers can quickly be compared and mapped in relation to those already discovered.

2.2.5 Variable number tandem repeats (VNTRs)

Restriction enzymes can also be used to detect polymorphism that is due to differences in the length of blocks of repeated DNA segments; that is, **variable number tandem repeats** (**VNTRs**).

Sometimes a short DNA sequence is present in several copies which lie one after the other forming a **tandem array**. The number of repeat units in tandem arrays is highly variable. This alters the length of fragments cut with restriction enzymes and produces many alleles at such a locus (Figure 2.11). This in turn leads to a high level of heterozygosity in the population; just what is needed for linkage mapping. The fragments are detected, like RFLPs, after gel separation, Southern blotting and hybridization with a radiolabelled probe.

In the example shown in Figure 2.12 each individual has two bands (i.e. two DNA fragments of different sizes), one from each copy of chromosome 16. Using this information it is possible to determine which chromosome 16 each of the children has inherited from its mother and which from its father.

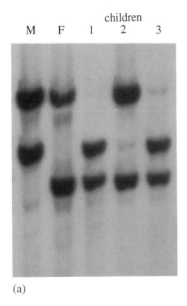

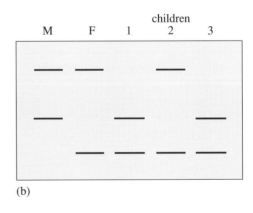

Figure 2.12 (a) Autoradiograph of a VNTR on chromosome 16 for a mother (M), father (F) and their three children (1–3). (b) Simplified drawing of the autoradiograph in (a).

▷ Draw out the relative sizes of this VNTR region from each of the two homologous chromosomes 16 of the mother and the father and work out which one has been inherited from each parent in each of the three children.

▶ We don't know the actual number of repeats along each of the four chromosomes 16, but from the relative positions of the bands on the autoradiograph, we know the relative sizes of the fragments. Denoting the four chromosomes as 16a–16d, the relative sizes of the fragments from the maternal and paternal chromosomes are as in Figure 2.13.

Thus the inheritance of the chromosome 16 in the children is as follows:

child 1: 16b from mother, 16d from father

child 2: 16a from mother, 16d from father

child 3: 16b from mother, 16d from father.

▷ If children 1 and 3 are affected by the same genetic disease as their mother, what does this tell you?

▶ As the alleles of the VNTR co-segregate with the disease it is possible that the VNTR locus is linked to the disease locus (and that the mutated copy of the 'disease gene' is on chromosome homologue 16b). However, it would be necessary to test many more families before linkage is proved.

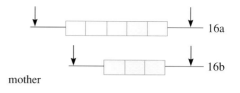

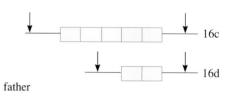

Figure 2.13 A VNTR on chromosome 16 (16a and 16b, maternal; 16c and 16d, paternal).

Usually at an RFLP based on a simple restriction site presence or absence the DNA is either digested or not, giving two alleles (see Figure 2.8). The maximum heterozygosity at such a locus is 50% (Box 1.1).

The heterozygosity of a locus will be increased if there are more than two alleles. In a three-allele system with alleles A, B and C, an individual can be heterozygous in three ways; AB, AC and BC.

The frequency of heterozygotes is calculated from the frequency of each allele.

The frequency of allele $A = p$, $B = q$, $C = r$

An extension of the Hardy–Weinberg formula for three alleles gives the frequencies of the six possible genotypes as

$$p^2 + q^2 + r^2 + 2pq + 2pr + 2qr = 1$$

and the frequency of heterozygotes as

$$2pq + 2pr + 2qr \quad \text{or} \quad 1 - (p^2 + q^2 + r^2)$$

▷ What is the heterozygosity at a locus with three alleles A, B and C where the frequency of each allele is (a) 0.1, 0.4 and 0.5 respectively and (b) 0.25, 0.25 and 0.5 respectively?

▶ (a) $p = 0.1$, $q = 0.4$, $r = 0.5$. Heterozygosity $= (2 \times 0.1 \times 0.4) + (2 \times 0.1 \times 0.5) + (2 \times 0.4 \times 0.5) = 0.58$ or 58%.

(b) $p = 0.25$, $q = 0.25$, $r = 0.5$. Heterozygosity $= 0.625$ or 62.5%.

As the number of alleles at a locus increases, higher and higher values of heterozygosity are possible.

The **minisatellites** are an extreme example of the VNTR type of sequence arrangement. They are short (10–60 bp) DNA sequences which are each repeated 10–100 times in tandem arrays. The number of repeat units in each block is very variable. Different alleles can be detected by digesting DNA with a restriction enzyme that recognizes a sequence outside the repeat. As described above, different-size fragments will result from individuals with different numbers of the repeated unit (Figure 2.11). They are detected in the usual way by gel electrophoresis and hybridization to a radiolabelled probe: in this case the probe is made from a sequence complementary to one of the repeat units.

Some VNTR loci have more than 10 alleles and the overall heterozygosity may exceed 90%. VNTR loci are thus very useful as markers for linkage mapping and also in forensic studies (Box 2.3).

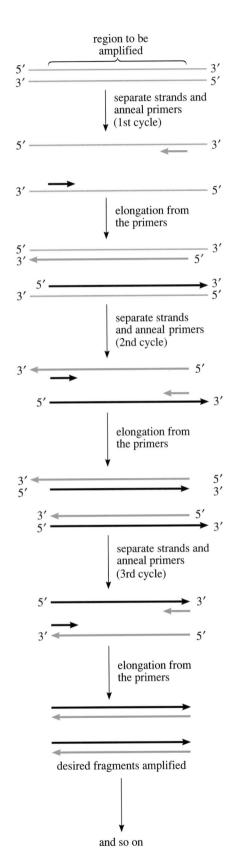

Figure 2.14 The polymerase chain reaction.

Box 2.3 Forensic applications of VNTR

By using a probe that hybridizes to many minisatellite regions simultaneously it is possible to identify several VNTR loci at once. Look again at the autoradiograph of alleles at a VNTR locus on chromosome 16 (Figure 2.12). Imagine an autoradiograph that showed the alleles at 20–30 loci at once. This gives a complicated pattern.

The number of alleles at each VNTR locus in the population is so great that the pattern of bands – the **DNA fingerprint** – is unique to each individual (excluding identical twins). Relatives share some bands depending on their degree of relatedness.

DNA fingerprints have been much used in detective work. DNA obtained from samples of blood or semen left at the scene of a crime can be compared to the DNA of suspects. There are questions concerning the reliability of such tests (the most cogent of which are concerned with the problem of shared fingerprints in populations in which many people are related), but they have proved useful in *excluding* individuals from a crime. They are also used in cases of disputed paternity by analysing DNA from mother, child and potential fathers.

2.2.6 Dinucleotide repeats – microsatellites

As we have seen, minisatellites involve repeated blocks of sequences that are tens of nucleotides long. There may also be differences in the number of tandem repeats of just two nucleotides (such as CA or GT). These dinucleotides may be repeated from 15–60 times in different individuals. Like minisatellites, these so-called **microsatellite** sequences occur frequently in the human genome – on average about once every 30–60 kb. Since many are polymorphic for the repeat number, they provide an additional source of marker loci usable in linkage analysis.

As they vary only by multiples of two nucleotides, the size difference between the alleles is too small to be detected by restriction enzyme digestion and agarose gel electrophoresis. Such polymorphisms are usually detected by another technique, the **polymerase chain reaction (PCR)** (Box 2.4), a method which has had a revolutionary impact on many areas of molecular genetics. As the name suggests, the PCR technique produces multiple copies of specific regions from small amounts of DNA, sufficient to be used in subsequent analysis.

When the region of DNA including the dinucleotide repeat has been amplified by PCR it can be separated on the basis of size by polyacrylamide gel electrophoresis (PAGE) – polyacrylamide is able to separate molecules with small size differences much better than is agarose.

Polymorphism is also found in the numbers of copies of three – or four – nucleotide repeat units. As well as their usefulness as markers for linkage mapping, trinucleotide repeats are associated with mutations leading to a number of genetic diseases (Chapter 4).

Box 2.4 The polymerase chain reaction

The polymerase chain reaction (PCR) is an *in vitro* technique widely used to amplify small amounts of DNA. The sequence of steps is shown in Figure 2.14. (See also *Source Book.*)

2.2.7   Localizing disease genes using polymorphic markers

The discovery of the new types of polymorphic sequences described above – even though we have no idea what (if anything) their function is in the genome – has provided thousands of markers for genetic mapping. No longer do we need to rely on the few loci determining blood groups or other phenotypically detectable polymorphic characters in order to map the position of disease-causing loci. Linkage analysis can show whether two loci are close to one another on the chromosome. This information is important when trying to track down genes involved in disease.

If some members of a family are affected by a genetic disease resulting from an hitherto unmapped gene, it is sometimes possible to identify marker loci which co-segregate with the disease. They must either be within the disease gene itself or closely linked to it.

▷   Look back at Figure 2.7. Although the two families show linkage of nail–patella syndrome to the ABO blood group locus, one shows perfect linkage of the marker trait to the disease and the other does not. Which one shows perfect linkage? What do the results suggest about how linkage analysis should be carried out?

▶   Family (a) shows perfect linkage; family (b) does not (remember individuals III-2 and III-3 who have the *O* allele but not the disease). The best situation would be to study many individuals in one family, and for some diseases single kindreds containing hundreds of individuals have been studied. Usually, to get enough data to be sure of the results, a large number of families is needed. Results from a single family can hence be misleading: if only family (a) had been used we might be tempted to conclude that the ABO blood group locus is extremely close to that for nail–patella syndrome, which it is not. In mapping loci, the CEPH database makes it possible for the results from a large number of families to be combined, giving more reliable data.

▷   There is an important assumption here – can you think what it is?

▶   We are assuming that the disease is always caused by the same gene. This is not always the case, as sometimes mutations in two different genes give the same symptoms.

Once a marker is found to be linked to a disease there is a starting point from which to look for the gene itself. The marker acts like a flag in the DNA sequence, pointing to the region where the *disease gene* (that is, the gene which, when altered, in some way leads to a disease phenotype) may be. The process is more efficient if **flanking** markers around the crucial gene are discovered (Box 2.5). Two marker loci are found, as close as possible on either side of the gene. This gives boundaries to the region of DNA that must be isolated.

Box 2.5   Finding flanking markers

Suppose that a disease shows 2% recombination with marker A and 3% recombination with marker B. We can arbitrarily place locus A on one side of the disease locus (Figure 2.15a). Locus B may be positioned in two places, either of which will produce 3% recombination with the disease locus (Figure 2.15b). To determine whether locus B is on the opposite side of the disease locus to A (making A and B flanking markers), the recombination between loci A and B can be examined. For the arrangement shown in Figure 2.15c, the observed recombination between A and B will be 1%; for arrangement 2.15d it will be 5%.

Note that, in many cases, the order and distance of markers will be known from previous genetic mapping studies and this method will not be necessary. These days, once a marker for a particular chromosomal region has been shown to be linked to the disease, published genetic maps can be consulted for other markers to try.

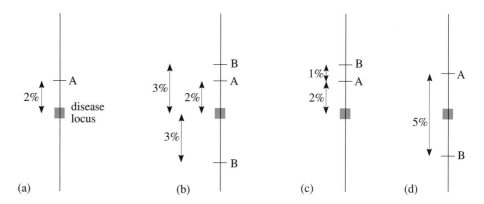

Figure 2.15 Analysis of recombination between a disease locus and markers A and B.

To find flanking markers many different probes may have to be tested on families with affected members. If the chromosome involved is not known, then markers from all chromosomes have to be used initially to find ones that identify on which chromosome the disease gene resides. Once this is known, more markers on that chromosome can be tested to find ones more closely linked to the disease gene.

Since some of the markers may not be useful or 'informative' in certain families as the affected parent may not be doubly heterozygous (Section 2.2.1), it is important to obtain DNA from as many family members and as many families as possible when carrying out a linkage study.

As the number of mapped DNA markers (from CEPH and other groups) increases the whole genome should become saturated with accurately placed landmarks. It will then be much easier to locate new disease genes, using sophisticated technology, but with an intellectual approach which is not very different from that introduced by Morgan and his co-workers at the beginning of the century. Finding markers that are closely linked to a disease gene does not mean that the gene itself is identified; to do this we must turn to other methodologies.

2.3   The triumph of technology: physical mapping

To build a genetic map based on linkage it is not necessary to examine the DNA directly: think, for example, of the linkage between the ABO blood group locus and that for nail–patella syndrome. **Physical** mapping, however, involves the direct determination, by chemical methods, of the positions of DNA segments on chromosomes. It depends on the manipulation of the DNA in various ways.

The two strategies, physical and genetic mapping, are often used in conjunction with each other to localize and isolate genes. Marker sequences close to the gene of interest are first identified by family studies as discussed above. The physical position of these sequences determines the region of the genome to be analysed in greater detail until the gene itself is located. What follows is a description of some of the ways that sequences can be localized to regions of the genome or chromosome.

2.3.1 In situ hybridization

If a marker whose chromosomal location is not known is found, by family studies, to be linked to a disease-causing gene, its physical position can be determined by hybridization to human chromosomes. A cell suspension is prepared, and spread onto a microscope slide. The chromosomes can be identified using standard staining techniques. The DNA on the slide is then denatured; the strands are separated making each one available for binding (hybridization) to a complementary sequence. The probe DNA (encoding the gene or sequence to be localized) is labelled and is itself denatured. The probe is then hybridized to the DNA on the slide. The position at which the probe binds indicates the location of its matching DNA.

There are various ways in which this position can be established. Initially, probes were labelled by incorporating tritium, ^{3}H, but this isotope has weak emissions and it takes a long time for them to be detected on photographic film. Most recently, probes have been labelled by the incorporation of bases attached to biotin. The biotin is detected via streptavidin (derived from avidin, the egg-white protein with a high affinity for biotin) which is conjugated with a fluorescent chemical. This procedure is termed **fluorescence *in situ* hybridization** (**FISH**) and is a fast and efficient way of localizing DNA sequences (Plate 2.1). This technique is most effective when large pieces of DNA are used as probes.

▷ Look at Plate 2.1. Notice that there are usually two spots on each metaphase chromosome (actually, one of these pairs has only one visible spot). Why is this?

▶ The metaphase chromosomes are dividing so each of the two copies of chromosome 3 consists of two chromatids. There is one signal on each chromatid.

▷ What is the large round blob with two yellow signals in it?

▶ This is a nucleus in interphase. At this stage the chromosomes are not condensed and so cannot be distinguished.

2.3.2 Somatic cell hybrids

The use of somatic cell hybrids is another way to find the chromosomal location of a DNA sequence.

It is possible to fuse cells of two different species (human and mouse or hamster, for example) under certain conditions (Box 2.6). When this is done, the daughter cells preferentially retain the mouse genome and, at each cell division, human chromosomes are lost at random. In different hybrid cell lines, different subsets of human chromosomes are lost. A 'panel' of cell lines which have retained different complements of human chromosomes (identified using markers whose positions are already known) can then be made. Such panels are very useful in localizing genes – either by testing for their protein products or, in the case of

DNA markers, by hybridization to a Southern blot (Box 2.2) or by PCR (Box 2.4). If hybrid panels are made from cell lines which contain deleted human chromosomes (see below) mapping can then be refined to subregions of a chromosome.

▷ Five hybrid cell lines contain human chromosomes as indicated below. If the marker DNA is shown by hybridization to a Southern blot to be present in cell lines A and C but absent from B, D and E, can you identify on which chromosome that sequence is located?

Cell line	Human chromosomes present
A	1, 5, 8, 10, 21
B	1, 5, 10, 12
C	1, 8, 12, 14
D	5, 14, 18
E	1, 3, 6, 7, 9, 15

▶ The marker is located on chromosome 8, the only chromosome unique to lines A and C. This example is a simple one, but with many panels containing different combinations of human chromosomes (or even parts of chromosomes) somatic cell hybridization can be a very effective mapping method.

Box 2.6 Cell fusion for gene mapping

The fusion of two cells can be achieved by the addition of Sendai virus or polyethylene glycol. In Figure 2.16 human and rodent cells are fused when a virus particle binds to two adjacent cells. A cell with two nuclei is formed. After cell division, the two daughter cells contain chromosomes from both parental cell lines. There will be a proportion of human–human and rodent–rodent cell fusions and the daughter cells from these are selected against (killed off) by the addition of various chemicals to the growth medium.

Usually the human chromosomes are lost as the hybrid cells continue to divide. (It is not known why this happens.) This is the basis for the gene mapping strategy.

Stable hybrid cell lines with varying chromosome complements can be examined for the presence or absence of marker DNA. This can then be correlated with a particular chromosome.

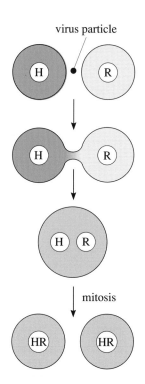

Figure 2.16 Fusion of mammalian cells induced by inactivated Sendai virus. H = human genome; R = rodent genome.

2.3.3 Deletion mapping

Deletion mapping depends on the fact that a particular disease phenotype can sometimes be correlated with a microscopically visible loss of a chromosome region. Deletions of chromosome regions can pinpoint the position of genes if they can be correlated with a particular disorder.

The positions of several genes have been determined in this way. A microscopically visible example is the loss of part of the short arm of chromosome 5. This results in **cri du chat** syndrome (see Chapter 4). A second good example is the localization of a gene involved in male fertility. It is known that infertility in males is sometimes associated with a deletion of a substantial piece of the long arm of the Y chromosome. A small number of such infertile (**azoospermic**) males lack just one very small portion of the chromosome. This deleted region is likely to contain one or more of the genes essential for sperm production. This region of the chromosome has been **cloned** (Box 2.7) and examined and found to contain a gene that may be involved in male fertility.

Some chromosomal deletions lead to a variety of apparently unrelated phenotypic effects; not surprisingly, the larger the deletion, the more body systems are involved. Such **contiguous deletion syndromes** are useful in localizing disease-causing genes to a chromosomal region. Genes which, when deleted, lead to each syndrome can be assigned to particular sites by correlating the details of the karyotype with particular clinical features. Figure 2.17 shows how increasing deletions of the short arm of the X chromosome result in an increased number of disease phenotypes. Patient A is short. Patient B, with the larger deletion, is both short and has X-linked recessive chondrodysplasia punctata: the chondrodysplasias take many forms but one feature of new-born affected individuals is that they have abnormal bone growth.

2.3.4 Recombinant DNA technology

Once a gene is assigned to a chromosomal region using one or more of the methods described above it is necessary to map it more precisely. The usual strategy is to cut the DNA and then clone it into DNA molecules called **vectors** which allow the selected portions to be amplified (multiplied in number) in bacterial or yeast cells. Cloning is illustrated in Box 2.7.

Vectors act as a vehicle for DNA. They allow it to be perpetuated and then to be extracted for analysis. Different vectors maintain different lengths of donor DNA (Box 2.8). It is a good idea to study this box closely as some of the different vector types will be referred to in the text that follows.

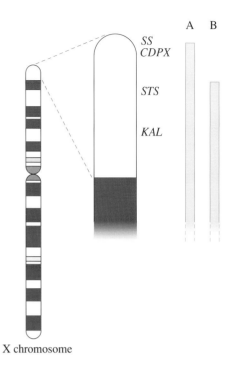

X chromosome

Figure 2.17 Some of the deletions from the end of the short arm of the X chromosome. The bands indicate the relative lengths of the regions remaining. Abbreviations for disease loci: *SS*, short stature; *CDPX*, X-linked chondrodysplasia punctata; *STS*, steroid sulphatase; *KAL*, Kallman syndrome.

Box 2.7 DNA cloning

The procedure used for gene cloning in *E. coli* is illustrated in Figure 2.18; the stages are described below.

(a) Plasmids (Box 2.8) are isolated from a pure culture of *E. coli*; that is, all the cells derive from a single ancestor. So all the plasmids isolated are identical.

(b) The plasmid circles are treated with a restriction enzyme which cleaves each plasmid at the *same* single site thus opening up each circle in the same way.

(c) DNA is extracted from some (foreign) cells.

(d) The molecules of DNA are cleaved into pieces with a restriction enzyme. This results in many *different* pieces of DNA.

(e) The pieces of the foreign DNA are spliced into the opened plasmid circles which are re-sealed to create *recombinant* plasmids. Many *different* recombinant plasmids are thus created, because the pieces of foreign DNA were different.

(f) When mixed with some plasmid-free *E. coli*, in the presence of calcium ions, individual recombinant plasmids are taken up by individual cells – usually one plasmid per cell, as the process is very inefficient.

(g) In each cell containing a recombinant plasmid, that plasmid replicates.

(h) Each cell grows, its DNA replicates (chromosomal *and* plasmid), and the cell divides; its offspring grow, replicate their DNA, divide, and so on. So at each cell division, a cell passes on a copy of its chromosomal DNA *and* identical copies *of its own particular recombinant plasmid* to its descendent cells. Thus *each* cell resulting from step (f) produces a clone of (identical) cells. The clones differ from each other according to what particular recombinant plasmid the cells carry; but *within* a clone, *all* the cells contains copies of the same plasmid. Among these clones is one (or perhaps more) where the recombinant plasmid contains the desired foreign gene.

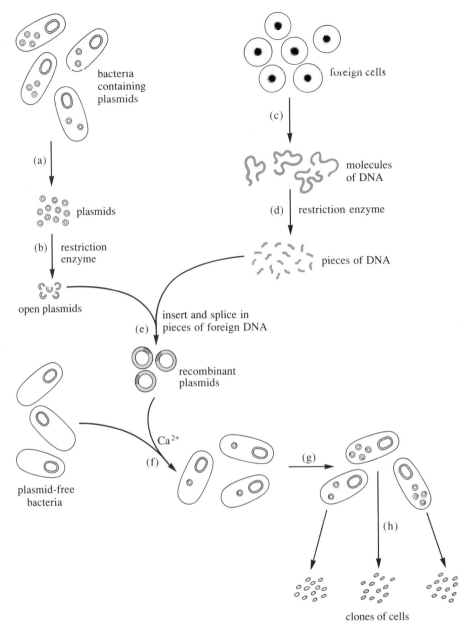

Figure 2.18 Outline protocol for DNA cloning. (*E. coli* DNA – chromosomal and plasmid – is shown in black; and foreign DNA in colour.)

Any vector must contain unique sites for a variety of restriction enzymes into which the donor DNA may be inserted. The **recombinant** molecule is then amplified in a bacterium (usually *E. coli*) using origins of replication present in the vector. In addition, vectors are engineered to confer antibiotic resistance so that when *E. coli* cultures are grown in media containing an antibiotic only those cells containing the vector, and the sequence of interest, will survive.

As described in Box 2.7, the first step in cloning is to cut up the DNA using restriction enzymes. The commonly used restriction enzymes cut genomic DNA in many places, producing many small fragments (usually around 1–4 kb long). The 3 000 million base pair human genome will be cut into around 3 million pieces, the diploid genome producing around 6 million pieces per cell.

Box 2.8 *Different types of vector*

The characteristics of six vector types are summarized below.

Plasmid

A plasmid is an extrachromosomal circular DNA molecule found in bacteria, and can accept foreign DNA fragments as described in Figure 2.18 of around 10 kb in size (Figure 2.19a).

Bacteriophage (phage) lambda

Phage lambda (λ) is a virus that infects bacteria and replicates inside the bacterial cell. In such infection, once inside a cell the genome of a single phage 'hijacks' the cell machinery causing the cell to make proteins coded for by *phage* genes and to replicate *phage* DNA. Ultimately, the newly synthesized phage proteins and molecules

of phage DNA assemble together to give many new phages. These then rupture the cell, freeing them to invade and infect many more surrounding cells, and so on (Figure 2.19b). Part of the λ genome can be used as a vector in a similar way to plasmids; it can accept inserts of 25–30 kb.

Cosmid

This type of vector is a phage–plasmid hybrid. The advantage of using cosmids is that they can support larger foreign DNA inserts – about 40 kb.

P1

A P1 vector is based on phage, but after injection of its DNA into the bacterial host it replicates as a plasmid (foreign DNA insert size 80–100 kb).

YAC (yeast artificial chromosome)

A YAC will hold a very much larger insert than the other vectors – up to 1 000 kb. As the name suggests, it is an artificial chromosome grown in yeast cells. It contains the three essential components of a chromosome: a centromere and two **telomeres** (ends). In addition there is a site of origin for DNA replication, the ARS (autonomously replicating sequence). The donor DNA is inserted between one telomere and the centromere (Figure 2.19c).

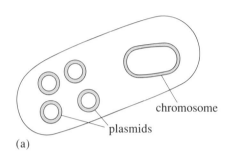

(a)

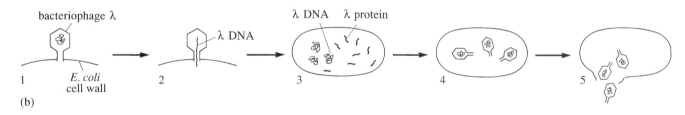

(b)

(c)

Figure 2.19 Examples of vectors. (a) Plasmid. (b) Bacteriophage. (c) YAC (yeast artificial chromosome).

To increase the size and hence reduce the number of pieces, restriction enzymes that cut at very few sites can be used. These 'rare-cutters' cleave the DNA into large sections, around 1 Mb (megabase = 10^6 bases) in size. The human genome then gives around 3 000 pieces.

Such lengths must then be cloned into the YAC vectors described in Box 2.8.

Each vector molecule, whatever vector is used, takes up a different piece of donor DNA. The collection of clones containing all the pieces is called a **library**.

The next task is to find which clone in the library contains the DNA sequence of interest – that is, the colonies must be **screened** (Box 2.9).

Box 2.9 Colony screening

Colony screening involves making an exact copy, or *replica*, of the colonies. As for RFLP detection (Box 2.2) this involves transfer of the DNA to a nitrocellulose filter. The colonies to be screened are grown on nutrient jelly in a Petri dish (the master plate, Figure 2.20a). A nitrocellulose filter disc is then brought into 'face-to-face' contact with the surface of the master plate – that is, the two surfaces, colony-bearing one and nitrocellulose disc, are lightly pressed together (Figure 2.20b). This transfers a small portion of each colony onto the nitrocellulose disc – and the array of the portions transferred onto nitrocellulose matches that of the colonies on the original dish. The disc is then removed (Figure 2.20c) giving a replica, on the disc, of the master plate. The transferred portions of clones, on this replica, can then themselves be grown by placing the disc on fresh nutrient jelly. All destructive procedures (e.g. extracting DNA) can then be done using the replica on the nitrocellulose disc (Figure 2.20d), retaining the master plate intact.

The replica is then treated with sodium hydroxide, a strong alkali. This has two effects. First, it breaks open the cells, thus exposing all their DNA (chromosomal and recombinant plasmid). Secondly, it causes the individual strands in each double-helical molecule of DNA to separate from each other, resulting in single-stranded DNA. The disc is then treated to remove protein but leaving the single-stranded DNA from each colony bound to the disc at exactly the same sites as were the colonies

themselves. Finally, the disc is baked at 80 °C, thus fixing the single-stranded DNA firmly in place (Figure 2.20e). As comparing (e) with (d) in the figure will show, this leaves us with a nitrocellulose disc bearing the 'DNA prints' (not to be confused with DNA fingerprints!) of each of the colonies present on the replica and, hence, present still on the master plate (Figure 2.20a). It is

these DNA prints which are now tested – that is, *probed* (as described in Box 2.2) – to determine which one contains the sequence of interest. Knowing the position on the replica filter of the clone containing the marker sequence means that the original colony can be identified and used in the next stages of the gene-hunting process.

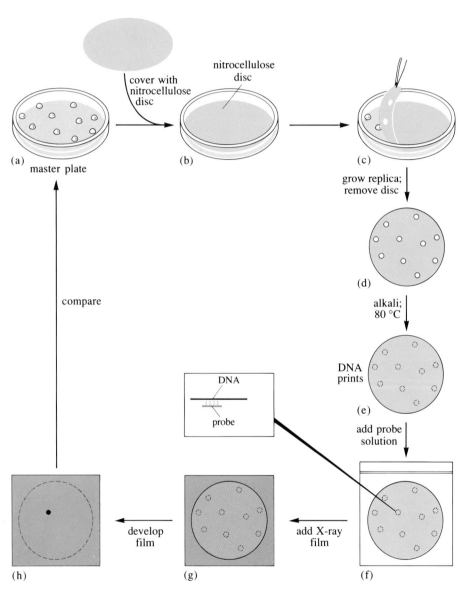

Figure 2.20 Outline protocol for colony screening.

The strategy of finding the chromosomal location of a gene, then cutting up the chromosome into fairly large pieces which are cloned, gradually closing in on the gene is termed a '**top-down strategy**'. However, it is possible to reconstruct a chromosome from much smaller DNA fragments without prior knowledge of where a particular gene might be. This is called a '**bottom-up strategy**' and has advantages of its own.

2.3.5 Bottom-up mapping

To reconstruct a segment of DNA it is necessary to produce a library of clones as described above. They may be prepared in plasmid, phage, cosmid or YAC vectors, depending on the size required. The problem is to understand how the clones fit together, rather like doing a jigsaw puzzle.

Bottom-up mapping depends on partial cleavage of DNA. If DNA is all cut completely by a particular enzyme there will be a series of sequential non-overlapping fragments (Figure 2.21a). However, if the DNA is cleaved for a shorter period then the enzyme will not have sufficient time to cut all possible sites – resulting in a **partial digest**. Many copies of the genome – many cells – are used in each partial digest experiment. This results in a series of overlapping fragments which vary in size, depending on where the enzyme happens to cut the DNA (Figure 2.21b).

Fragments emerging from a partial digest contain some regions of homology (overlap) with each other and some regions that are not in common. By determining which fragments have regions in common it is possible to infer which fragments overlap with which. A series of sequential overlapping clones can be constructed. A complete set of these will, eventually, cover the whole of the original material. Such a series of sequential overlapping fragments is called a **contig** since it represents *contiguous* sequences (joined in order) (Figure 2.21c).

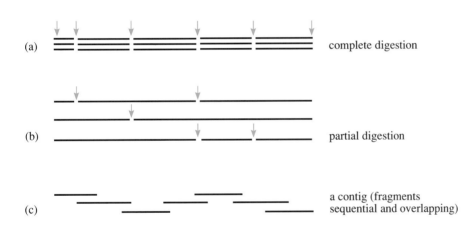

(a) complete digestion

(b) partial digestion

(c) a contig (fragments sequential and overlapping)

Figure 2.21 Partial digestion of DNA. (a) Result of *complete* digestion – multiple copies of non-overlapping fragments. (b) Result of *partial* digestion – overlapping fragments. (c) A contig: a series of overlapping cloned fragments which represent the starting material.

When large genomes are involved such as the human or mouse genome – or even the 8×10^7 bp genome of the nematode *Caenorhabditis elegans* on which the technique was developed (about the size of one human chromosome) – a large number of clones is required. To make the projects feasible, vectors that can tolerate quite large inserts are used, reducing the number of clones. Cosmid vectors (see Box 2.8) are often used, as they can support a 35–40 kb insert size.

To determine how the cosmid clones overlap, a form of restriction map is produced. Since restriction enzymes cleave the DNA at specific sites then the pattern of bands produced when the products (the restriction enzyme digests of each cosmid clone) are separated by electrophoresis is unique to each clone. This is often called a **cosmid fingerprint**. Clones that overlap have some DNA in common and share some of the same-size digestion products. If two clones have more than half their cut lengths in common then they must overlap. (Figure 2.22). Contig mapping involves making fingerprints for many clones which are then compared with each other, usually by computer.

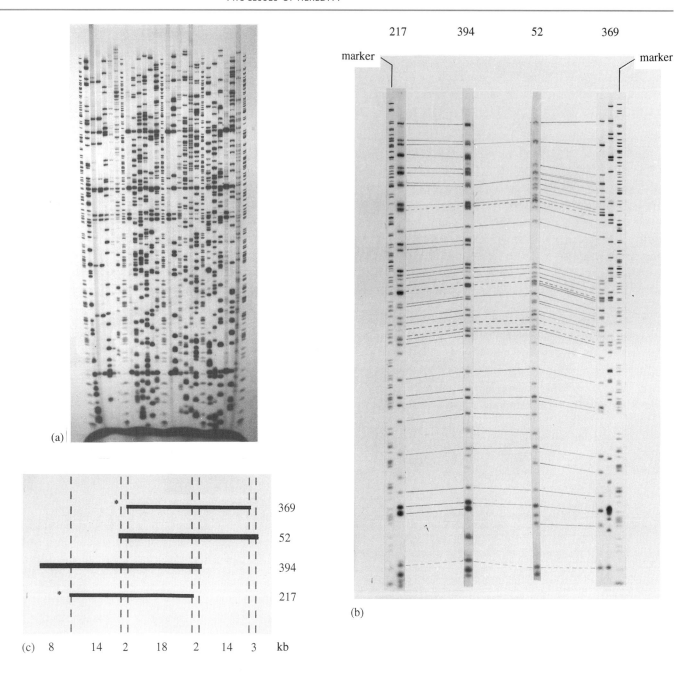

Figure 2.22 An example of bottom-up mapping: fingerprinting of cosmid clones of the human Y chromosome. (a) Photo of a cosmid fingerprinting gel. The gel contains fingerprints of 24 cosmid clones. Every seventh lane is a marker lane. (b) If you look at these four cosmid fingerprints in detail, you can see that they share bands in common (joined with continuous lines). Some of the bands are derived from the vector DNA and so are the same in *all* of the clones (these bands are joined with dashed lines). (c) This diagram shows how the four clones overlap with each other. In fact, the two clones marked with an asterisk have bands that are completely contained within the other two. Clones 52 and 394 have about 50% of their insert length in common, thus we can be confident that they overlap.

2.3.6   DNA sequencing

The finest level of resolution of a physical map is the nucleotide sequence itself. As the methodology used for sequencing can only at present be carried out on short cloned fragments, this is best done when the region of interest is narrowed down as much as possible. Overlapping clones from the regions described above can be used as a starting point. The most widely used procedure (the Sanger method – Figure 2.23) requires that larger cloned DNA molecules are broken down into fragments of around 400 bp and cloned into a vector (M13) from which a single-stranded template is produced. Using an oligonucleotide which matches the vector sequence next to the cloning site, a DNA polymerase produces a DNA strand complementary to the template. Four reactions are set up for each template. Each is supplied with the four deoxyribonucleotide triphosphates, dATP, dCTP, dGTP and dTTP, but in addition a small amount of 'poison pill' is included. In each of the four reactions this is a dideoxy version of one of the four deoxyribonucleotide triphosphates. Whenever these are added to the chain, the elongation reaction stops at the appropriate base. For example, in the reaction containing dideoxyadenosine triphosphate (ddATP), molecules of differing length all terminating in adenosine are made. As only a small amount of dideoxyribonucleotide triphosphate (ddNTP) is added, the reaction will consist of a series of elongation products which have terminated at different positions along the template. When the products of all four termination reactions are run alongside each other on a polyacrylamide gel, a 'ladder' of fragments of increasing length is produced. From this the nucleotide sequence can be read.

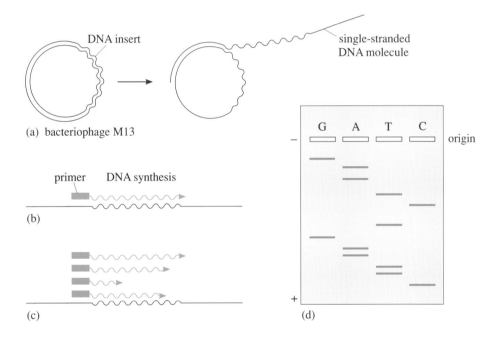

Figure 2.23 The Sanger method for DNA sequencing. (a) The DNA is cloned into a vector called M13. This is a bacteriophage that produces a single-stranded molecule at a certain stage in its life cycle. (b) With the addition of a primer sequence, the single-stranded DNA molecule is used as a template for DNA polymerase. (c) The reaction is carried out in four tubes each containing a mixture of the four dNTPs (deoxynucleotide triphosphates) which become incorporated into the growing chain. In each tube is a low concentration of just one of the dideoxynucleotide triphosphates (ddNTPs). The incorporation of one of these causes the chain to stop growing. Thus a series of fragments are produced which end in one of the four nucleotides (with the contents of each tube having a range of lengths but all ending with the same nucleotide). (d) The products from each tube are separated on an agarose gel and visualized by autoradiography (a small amount of radioactive nucleotide is included in each reaction). The DNA sequence is the read from the bottom of the gel upwards.

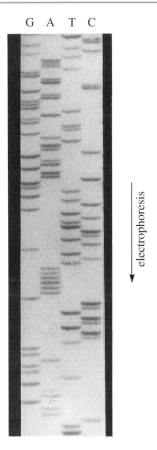

G A T C

electrophoresis

Figure 2.24 Photo of a DNA sequencing gel.

▷ You should be able to see that the sequence of the DNA in the example in Figure 2.23d is CTTAAGTCTAAG. Now read the sequence from the photograph of the real sequencing gel shown in Figure 2.24.

▶ The sequence is:

TTCAAGAAGTGGAGGTCCTCCTCCGAAAAAATCTGCTCCTT CTGCTGTGGCAAGAAGCAATAGTTGGATGGGAAGCCAAGG TAAATGCTGCCTGA

One aim of DNA sequencing is to locate – or map – functioning genes. Another is to sequence the DNA of whole genomes in order to examine the structure and function of the DNA that does not code for proteins (about 90% of the genome).

This is an enormous task which takes a lot of human resources and is also very expensive. Much work has gone into automating the method. Originally, as described above, termination products were radioactively labelled, separated by PAGE, visualized by autoradiography and the sequence read by eye. A more recent technique is to label the termination products with fluorescent molecules. Each termination reaction is labelled with a *different* fluorescent 'tag'. Fragments ending with each of the four nucleotides can then be distinguished from each other even when separated in the *same* lane of the same gel. The samples are allowed to run off the bottom of the gel and the labels are detected by a laser beam. A longer sequence can then be read from one reaction and also more reactions can be run on the same gel. Sequence recording is done by a computer which registers which fluorescent labels arrive at the bottom of the gel and in what order.

2.4 The genetic chart

The primary purpose of creating either a genetic or physical map of an organism is to localize and clone genes. This section describes how maps made in these two different ways can be compared.

Another reason for mapping is to study the bulk of the DNA which does not code for proteins. What does the vast amount of supposedly 'junk DNA' consist of, what functions, if any, can be assigned to it and how can this be searched through to find the genes? Some of these questions are addressed in this and the following section (2.5). Later, in Section 2.6, we shall discuss some of the methods that might be used to identify coding regions.

2.4.1 Comparing maps

For different maps to be compared they must use the same points of reference, or markers. St Paul's Cathedral is distinguishable both on an Ordnance Survey map of London and in an *A to Z*. In mapping chromosomes, it may be necessary to compare two maps constructed by different laboratories, or using different vectors.

A simple form of marker has been developed for this purpose – called a **sequence tagged site** or **STS**. A sequence tagged site is, as the name suggests, a short stretch of around 200 bp of DNA of known sequence. From this, unique oligonucleotides are designed to act as primers for the polymerase chain reaction (refer back to Box 2.4). The primer sequences are maintained in a computer database and often sold by commercial companies, so that any researcher can synthesize or purchase the primers and carry out the PCR reaction on their own clones. Hence different groups can localize exactly the same markers on a variety of maps.

In addition to genomic DNA, STSs can be designed to the sequence of cDNA clones. Remember that cDNA is a complementary DNA copy of RNA. Since cDNA clones therefore represent expressed DNA, the markers derived from them are called **expressed sequence tagged sites** or **ESTSs**. ESTS markers have been very useful in defining the map location of genes rather than the 90% of DNA that does not code for proteins. They are like a tourist map of a city which indicates just the most interesting buildings.

2.4.2 Genetic and physical maps are similar but not identical

If two maps of the same chromosome are constructed using the same markers but one using linkage mapping and the other using physical mapping methods, they will appear slightly different. This is demonstrated by the genetic and physical maps of yeast chromosome III (Figure 2.25).

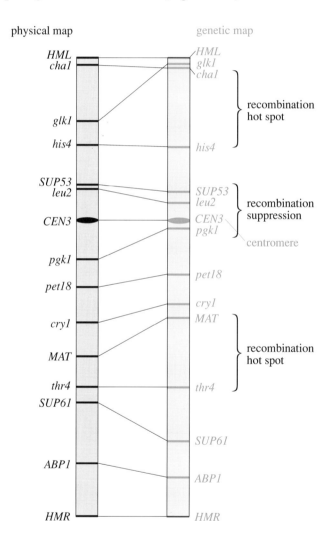

Figure 2.25 The physical and genetic maps of yeast chromosome III. The same markers can be seen on both maps but the relative distances between them vary depending on the method used to position them. Notice that markers at or near the centromere (*pgk1*, *CEN3* and *leu2*) appear closer together on the genetic map than the physical map, whereas the *MAT* and *thr4*, and *glk1* and *his4* loci are further apart on the genetic map.

The main reason for this discrepancy is that recombination rates vary at different chromosomal positions. At **recombination hot spots** the recombination rate is significantly higher than other regions, whereas some regions are reduced in recombination – there is **recombination suppression**. The nature of recombination hot spots has not yet been elucidated but there is evidence for such sequences in bulk DNA and even in the middle of genes. (For example, within an 18 kb region of the gene encoding the glycolytic enzyme phosphoglucomutase, the recombination rate is around 27 times the genome average.) Recombination suppression occurs in the vicinity of another recombination event, due to physical constraints, and also in regions of condensed chromatin called **heterochromatin**. There is a distinct reduction in the amount of recombination around the centromere and also at the telomeres.

▷ Why do recombination hot spots or recombination suppression alter the apparent distance between two markers?

▶ Since the number of recombination events is used as an indication of physical distance (1 recombination event per 100 meioses = 1 cM = 10^6 bp), a recombination hot spot will make the two markers seem further apart than they really are; recombination suppression will make them seem closer.

▷ What is the genetic map length (in centimorgans) in humans, given that the accepted size of the haploid genome is 3 billion (3×10^9) base pairs?

▶ Since 1 cM is approximately equivalent to 1×10^6 bp, the theoretical genetic map length is 3 000 cM.

The human **pseudoautosomal** region located on the short arms of the X and Y chromosomes has markedly higher recombination rates than the genome average. This region is the only portion of the sex chromosomes that undergoes pairing during meiosis – hence the name (it behaves like the autosomes). It appears that at least one chiasma (cross-over) must occur between the X and the Y chromosome to allow meiosis to work properly (Figure 2.26). The physical size of the pseudoautosomal region is around 2.3×10^6 bp, but there is 50% recombination between the sex-specific region and the telomere which would normally correspond to a size of 50×10^6 bp; thus, the rate of recombination in this region is more than 20 times as high as the genome average.

The interpretation of genetic maps is further complicated since recombination rates differ between males and females. Since the recombination rate is higher in female gamete formation, genetic maps made using data from female meioses are longer (3 850 cM compared to 2 750 cM in males). Published genetic maps are usually **sex averaged**.

2.4.3 The C-value paradox

When localizing genes to the map, perhaps using ESTS markers, it becomes very clear that the genes are few and far between. It is well known that the amount of DNA in the human and other eukaryotic genomes is far in excess of the amount required to encode [**make up?**] the genes.

The amount of DNA in a haploid genome is termed the ***C-value***. Among organisms, the *C-value* tends to increase with evolutionary complexity, as expected. However, in eukaryotes, the *C-value* is far greater than expected based on the estimated number of genes – this is referred to as the ***C-value* paradox** (Table 2.1). For example, the genome of the pea plant is one and a half times the size of the human genome.

In humans, the size of an average gene is about 20 000 bp and there are in the region of 50 000 genes (although estimates range from 30–100 000). The amount of DNA required to encode these genes is therefore $20\,000 \times 50\,000 = 1 \times 10^9$ bp. The genome is actually around 3×10^9 bp and so the genes account for only around one-third of this.

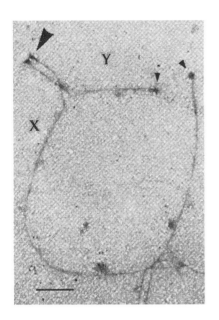

Figure 2.26 Pairing of the X and Y chromosomes at meiosis. A recombination event always occurs within the short pairing region (called the pseudoautosomal region) making this segment appear much larger on genetic maps than on physical ones. The large arrowhead is pointing to the X and Y short-arm telomeres and the small arrowheads to the long-arm telomeres.

Table 2.1 DNA content per haploid genome (*C-value*) of a range of organisms.

Organism	DNA quantity/bp
bacteriophage lambda	5.0×10^4
bacteriophage T4	1.7×10^5
E. coli bacterium	4.0×10^6
yeast	2.4×10^7
maize	3.9×10^9
lily	3.6×10^{10}
pea	4.5×10^9
Drosophila (fruit-fly)	1.8×10^8
salamander	8.4×10^{10}
mouse	2.5×10^9
human	3.0×10^9

So what does this additional DNA do? What is its function? Some of the excess DNA can be accounted for in terms of the regulatory sequences of the genes, or genes that have lost their function – the pseudogenes (introduced in Section 1.5.2). However, in contrast to the prokaryotic genomes where virtually all of the DNA is present only once and is called **single-copy**, much of the DNA in the eukaryotic genome is **repetitive** – present many times. We have already seen some examples of this in the minisatellites discussed in Section 2.2.5. In humans, these and other repetitive sequences account for about one-third of the total amount of DNA.

2.4.4 Reassociation kinetics

Not all types of repeated DNA are present at the same frequency; there are three approximate categories of DNA sequence: highly repeated, moderately repeated and single-copy.

The amounts of these three different forms in any one genome can be determined by carrying out experiments based on reassociation kinetics. Short lengths of genomic DNA denatured to form single-stranded molecules will anneal back together (or *reassociate*) under certain conditions. The amount of **reassociation** that occurs to form double-stranded molecules is a function of: (a) the *molecular concentration* of the sequence (sequences that are represented many times are more likely to encounter a complementary sequence) and (b) *time* (more molecules will reassociate if given more time to find their partner).

The amount of DNA present in the single- or double-stranded form can be determined by the absorption by the DNA of ultraviolet (UV) light: single-stranded DNA absorbs more UV light at a wavelength of 260 nm than does the same amount of double-stranded DNA.

The following describes the classic experiment by which the complexity of a sample of DNA may be determined. The original concentration (c_0) in moles per litre of a sample of DNA is calculated. This DNA is sheared and then denatured by heating to 100 °C, and the UV absorbance then measured. The sample is then cooled to 65 °C and maintained at this temperature. The proportion of DNA that has reannealed at various time points is determined by measuring the UV absorbance. A graph is plotted of absorbance (*y*-axis) against the logarithm of c_0 multiplied by the time in seconds (*t*) (*x*-axis). The resulting curve is called a $c_0 t$ curve (Figure 2.27) and consists of three different phases corresponding to the three classes of repetitive DNA.

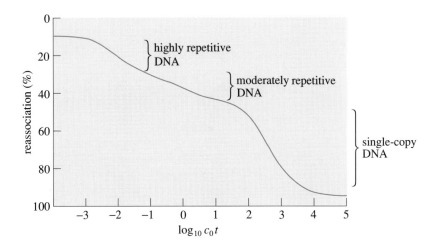

Figure 2.27 A $c_0 t$ curve – a profile of the molecular reassociation of human DNA.

2.4.5 Highly repeated sequences

The upper portion of the $c_0 t$ curve with the fastest rate of renaturation corresponds to **highly repeated** sequences which make up about 10% of the human genome. These may be repeated as many as 500 000 times (although the average is nearer 50 000) and can be present in **tandem arrays** (like minisatellites) or be **interspersed** among single-copy sequences.

Tandem arrays

The so-called *alphoid* sequences, first isolated in the African green monkey, are present in tandem arrays at the centromeres of human chromosomes. The basic unit is 171 bp which is repeated to form blocks up to several million base pairs in length. These sequences are not transcribed, but since the organization of the arrays is chromosome-specific, it is possible that they play some role in centromere function (Figure 2.28).

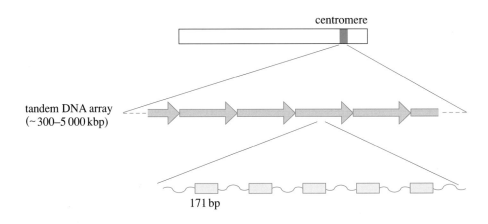

Figure 2.28 The organization of tandem repeats at the centromeres of human chromosomes.

Tandemly repeated sequences are also found on the long arm of the human Y chromosome, the heterochromatic region of which appears to be devoid of genes and to consist almost entirely of Y chromosome-specific tandemly repeated sequences. Men in a population are polymorphic for the size of their Y chromosomes due to the fact that these sequences may be reduced or increased in number with no apparent consequences. It is thought that the numbers of this type of repeated sequence can increase or decrease by a process called **unequal crossing over**. This is best explained with the aid of a diagram (Figure 2.29).

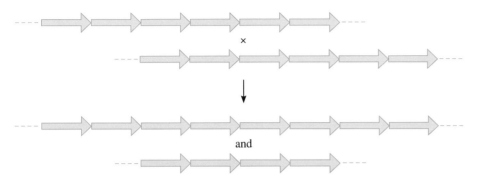

Figure 2.29 Unequal crossing over between homologous chromosomes with blocks of tandem repeats. (Only one chromatid of each chromosome is shown.) If, at meiosis, the chromosomes do not line up so that the repeats are exactly in register, a recombination event will have the effect of reducing the number of copies on one chromatid and increasing the number on the other.

At human telomeres six nucleotides, TTAGGG, are repeated over 1 000 times. It has been shown that the number of repeats decreases with the age of the cell and correlations have been made between the reducing number of these sequences and the control of cell death. The reverse has also been shown; telomeric repeats in cells that are immortal (i.e. cancer cells) are maintained in many copies by a cellular enzyme (telomerase).

Interspersed repeats

Interspersed repetitive sequences are described according to their length, there are Long INterspersed repetitive Elements (LINEs) or Short INterspersed repetitive Elements (SINEs). The 300 bp *Alu* sequence family is the most abundant example of SINEs in the human genome, with several hundred thousand copies. In contrast to sequences in tandem arrays, *Alu* elements are often transcribed and are present in nuclear pre-mRNA before splicing occurs (see Section 2.5).

LINEs are greater than 500 bp in length and, in the human genome, are represented predominantly by the L1 sequence which is 6 400 bp long and repeated from 3 000 to 40 000 times. They are not usually found in RNA species.

Alu sequences, which are G+C-rich, are present mainly in Giemsa-light bands and L1 sequences (A+T-rich) are mostly found in Giemsa-dark bands (Section 2.1.1). It is not yet known whether the distribution of these sequences accounts for the staining properties of the chromosomes.

2.4.6   Moderately repeated sequences

Moderately repeated sequences account for between 20 and 30% of the total human genome and they are repeated between 50 and 5 000 times – around 500 times on average. The most familiar examples are the genes for 28S and 18S ribosomal RNAs. There are several hundred copies of the ribosomal RNA genes which are clustered on the short arms of the **acrocentric** chromosomes – those with the centromere nearer one end than the other (Figure 2.6).

2.4.7   Gene families

Some sequences fall into a category which are neither single-copy nor middle repetitive. The genes encoding the transfer RNAs are an example of this. Each of the 60 different tRNA genes is present in around 10–20 copies.

None of the sequences described in this section so far code for a protein. The globin genes considered at more length in Chapter 1 are the classic examples of sequences present in several copies which *do* code for a series of proteins. Other examples of structural gene families include crystallin, the protein which makes up the bulk of the cell content of the fibre cells in the lens of the eye (Book 1, Chapter 3) and actin, the major component of muscle and contractile mechanisms in cells (Book 2, Chapter 7).

▷   Under what circumstances might it benefit an organism to have more than one copy of a gene?

▶   If a gene product is required in great quantities, then transcription from one gene copy may not be sufficient; increasing the number of templates would be a way around this problem. Alternatively, a second copy of a gene that is not essential in terms of producing sufficient mRNA is free to accumulate changes in its DNA sequence, some of which may be beneficial and confer an improved or alternative function on the protein. This seems to have occurred in the case of the globin genes.

Thus, around one-third of the genome of higher eukaryotes is composed of repeated sequences. The unique sequences tend to be interspersed between these repeated sequences, so that 300–600 bp of repeated DNA alternates with around 800–2 000 bp of unique sequences. The arrangement is different for different eukaryotes. In insects the tracts of repeated sequences are much longer. The reasons for this relationship between unique and repeated sequences can, as yet, only be speculated upon, but we can see that at least some of the 'junk' DNA plays vital roles in normal gene function. Some repeated sequences contain DNA needed for gene regulation (see Chapter 3). The repetitive sequences at telomeres are crucial for the structure of the chromosome, and those at centromeres may be essential for successful pairing at meiosis. Within genes are sections of non-coding DNA called *introns* (see Section 2.5). These are often conserved through evolution and thus are likely to play important roles in gene function. An exciting finding is that the genome of the pufferfish is about eight times smaller that expected and has much less 'junk'. This organism may help us identify the important regions within the non-coding portions of DNA.

2.5   The interrupted gene

We have seen in the previous section that, in marked contrast to prokaryotes, a proportion of the DNA of the eukaryotic genome seems surplus to requirements. The single-stranded circular genome of the phage φX174 is a good prokaryotic example. This short genome (5 375 bases) encodes nine different genes and an extremely small amount of the genome is non-coding DNA (only 4%). In two cases, the same DNA sequence is read twice, each time using a different reading frame(Figure 2.30).

In eukaryotes, even the genes themselves contain sections of DNA that do not code for parts of the protein.

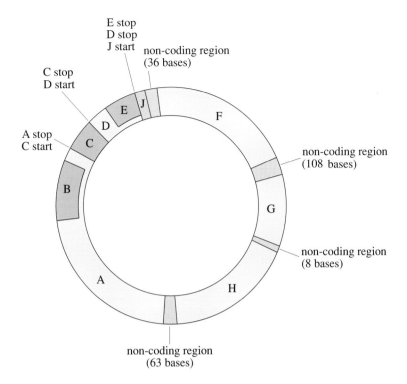

Figure 2.30 Arrangement of the nine genes found in the genome of phage φX174. Non-coding regions are unshaded. Notice that genes B and E occur entirely within other genes. Note also that the ends of some of the genes overlap by a few nucleotides with the starts of others.

2.5.1 Structure of expressed genes

As you know, for a protein to be produced from a DNA template, two processes must occur. The gene is **transcribed** to produce an RNA copy. The RNA is then used to produce the protein in a process called **translation**.

The portion of a gene that is transcribed is called the **transcription unit**. This includes **exons** and **introns**. The exons encode the amino acid sequence. The introns, or intervening sequences, are transcribed but not translated – the corresponding RNA is produced but this is not used to direct protein synthesis. The RNA copy of the complete transcription unit is termed **pre-messenger RNA** (**pre-mRNA**) or the **primary transcript**. Intronic sequences are removed or **spliced** (Chapter 3) out of this to form **messenger RNA** (**mRNA**) which is transported out of the nucleus to the cytosol where translation takes place (Figure 2.31).

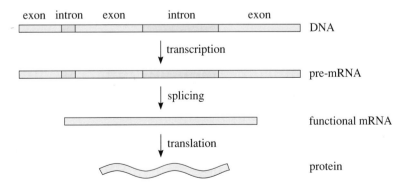

Figure 2.31 DNA is transcribed in the nucleus into pre-messenger RNA (pre-mRNA) or the primary transcript. The introns are removed from this to form messenger RNA (mRNA) which is transported into the cytosol where it is translated.

The presence of introns within genes was determined by a series of elegant experiments called **R-loop mapping** using the eukaryotic DNA viruses adenovirus and SV40 and later, with the human β-globin gene. The 15S pre-mRNA transcribed from the β-globin gene (referred to here as *nuclear precursor RNA*) is hybridized with cloned β-globin gene *DNA*. Under conditions where the DNA–DNA interactions are displaced by more stable DNA–RNA interactions, a loop of displaced single-stranded DNA is seen – one strand of the DNA pairs with the RNA, the other forms a so-called 'R-loop' (Figure 2.32a).

When the mature β-globin *mRNA* (9S) – introns removed – is hybridized to the same cloned DNA a loop of *double-stranded* DNA is produced in addition to single-stranded loops. Absence of intronic sequences from the mRNA leaves a region of double-stranded DNA corresponding to the intron of the gene. (Figure 2.32b). There are actually three exons and two introns in the β-globin gene but one of the introns is too small to be detected by this method.

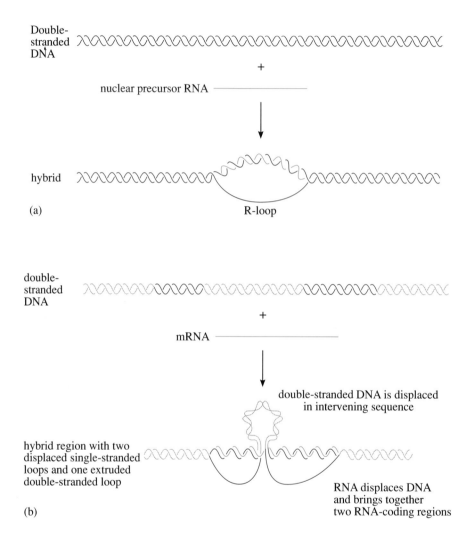

Figure 2.32 R-loop mapping.
(a) Under the appropriate conditions, RNA hybridizes with the corresponding gene by displacing one strand of the DNA double helix.
(b) When mature mRNA is hybridized with an interrupted gene, the intervening sequence (intron) cannot hybridize and remains as double-stranded DNA. (Exons are shown in black, introns in grey.)

The function of introns is not known. However, since introns are a feature of the genes of all eukaryotic organisms, it is most likely that they have some role to play in the process of regulation of gene expression. In general, introns of the same gene vary widely between species. However, some are remarkably conserved, suggesting that these particular introns have an important function.

2.5.2   Examples of gene structure

All members of the α- and β-globin gene families have three exons and two introns (Figure 2.33).

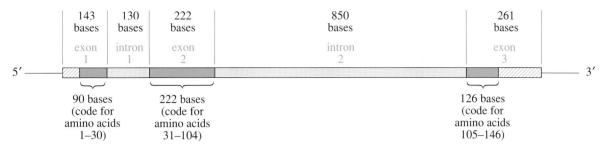

Figure 2.33 The structure of the β-globin gene. The exons (coloured bars) include the regions that encode the protein chain; the hatched segments represent regulatory regions that are, like introns, transcribed but not translated; these will be described in Chapter 3. The two introns are represented by grey bars.

Although the position and size of introns is conserved within a gene family, the sequence within introns varies substantially. On the other hand, the DNA sequence of exons of the same gene in different species is usually highly conserved. This makes sense since it is the exons which contain the information required to code for the protein. In general, exons are small, around 200 nucleotides on average. Some exons correspond to particular functional domains of the corresponding protein; for example, exon 2 of a globin gene encodes the haem-binding domain. However, not all exons can be correlated with functional domains in this way.

In humans, the number of introns within a gene can range from zero in the case of histone genes to greater than 50 in the collagen genes and at least 75 in the gene for dystrophin. (Dystrophin is a protein involved in muscle structure, and a deficiency results in Duchenne muscular dystrophy, an X-linked recessive condition: see Chapter 1.) In contrast to exons, introns vary dramatically in size – from 100 bp to over 100 000 bp. On average, there is 10 times more intron DNA than exon DNA in mammals.

The β-globin gene encodes 146 amino acids and spans around 1 500 bp of DNA. Thus around one-third (3 × 146 = 438 bp) of the gene is required for protein coding. The myoglobin protein, which functions as an oxygen carrier in muscle, comprises 155 amino acids, and is similar in size to β-globin. The gene that encodes myoglobin, however, is over 10 000 bp long and contains two introns, of 5 800 and 3 600 bp. Thus the coding sequences make up only around 5% of this gene – a strikingly small amount when compared to the genome of φX174 (with 96% coding DNA). A further example of a small protein encoded by a relatively large gene is hypoxanthine-guanine phosphoribosyl-transferase (HPRT) – an enzyme involved in purine metabolism. A partial deficiency of HPRT leads to gout and a total deficiency leads to Lesch–Nyhan syndrome, a condition characterized by severe physical and neurological abnormalities. The protein is 218 amino acids long and is encoded by nine exons which together make up less than 2% of the whole gene (42–44 000 bp) ; the average intron size is 5 000 bp. The largest human gene so far described is that encoding dystrophin. The protein is 4 000 amino acids long, which is about 10 times as large as the average protein. The coding region of this gene is comprised of 75 exons covering 2 000 kb, of which, the coding portion is only 1%. These data are summarized in Table 2.2.

Table 2.2 Examples of proportion of coding DNA within eukaryotic genes.

Gene	No. of base pairs	No. of amino acids in protein	% coding DNA
β-globin	1 500	146	33
myoglobin	10 000	155	5
HPRT	42–44 000	218	2
dystrophin	2 000 000	4 000	1

2.5.3   Pseudogenes

As you know, pseudogenes are sequences that are closely related to functional genes but are incapable of coding for the normal gene product. The reason for this is due to the presence of one or more mutations that result either in complete lack of transcription or transcription of an RNA molecule that is not translated in full. The pseudogenes of the α- and β-globin gene clusters in humans have several differences from the normal gene copy, but the ψ ζ (pseudo-zeta) gene (in the α cluster) has a single base change creating a 'stop' codon which terminates translation at amino acid 7, resulting in a non-functional protein.

There are two classes of pseudogene:

(a) Those that have arisen by mutation of a duplicated copy of the functional gene. The globin pseudogenes, discussed above, are examples of this class.

(b) Members of the second class of pseudogenes contain the exons of the functional gene but not the introns, and are usually not located near the functional copy. It is thought that these sequences have arisen from a messenger RNA – further evidence for this is that they often possess a poly A tail (Chapter 3). It appears that a processed RNA molecule (mRNA) is occasionally copied into a cDNA molecule by an enzyme called reverse transcriptase. This enzyme is an important feature of the life cycle of certain viruses called **retroviruses**, which copy their own RNA genome into cDNA so that it can be integrated (by a second enzyme, **integrase**) and perpetuated in the genome of the host. These two enzymes may be supplied in a mammalian germ cell by a virus and enable a DNA copy of an mRNA to be inserted randomly in the genome. Alternatively, the enzymes may be supplied by *transposable elements* (discussed in Chapter 4) which exist in the human genome. The random integration of such **processed pseudogenes** into the genome is likely to be selected against since integration may adversely affect an existing gene.

2.5.4   Conclusion

It was stated earlier that genes (including introns) are coded for by around one-third of the total DNA, the remainder being repeated sequences of one sort or another. When the presence of introns is taken into account, it is apparent that less than 5% of the total human DNA is likely to encode proteins (the average intron to exon ratio is about 1 : 10; that is, only about 10% of the gene contains the code for the protein). The low informational density of genes is an important point to bear in mind when searching through the genome to clone a particular gene.

2.6   Finding the coding regions

In this chapter we have described how DNA can be examined in order to identify the location of particular genes. The region of interest can then be cloned in a number of ways and analysed at a fine-scale level, allowing the identification of a particular gene. The large amount of repetitive DNA in eukaryotic genomes complicates this process since much of the material that is cloned does not code for proteins. It is necessary to search through this to find expressed sequences – rather like looking for a needle in a haystack. As we have said, the number of genes in the human genome is not known exactly but is in the region of 50 000–100 000 and these account for less than 5% of the total amount of DNA. However, there are features common to many genes which point the way to researchers trying to find an expressed sequence in a region of cloned DNA. Some or all of the features described below have been used to isolate particular disease genes. You will look at an example of this in Chapter 8, which deals with the identification of the cystic fibrosis gene.

2.6.1 Characteristics of expressed sequences

CpG islands

In Section 2.3 'rare-cutter' restriction enzymes were presented as a way in which DNA could be cut up into large fragments. In bulk DNA the cytosine of the CpG dinucleotide is often methylated – it has a methyl (CH_3) group attached to it (see Figure 5.4a in Book 1) – and this inhibits cleavage by restriction enzymes.

Methylation of cytosine nucleotides is thought to be an important aspect of the control of gene expression (see Chapter 3). At the 5′ ends of genes the majority of CpGs are not methylated (these regions are referred to as **hypomethylated**). These hypomethylated clusters of CpGs (**CpG islands**) are cleaved by the 'rare-cutter' enzymes. Hence cleavage of genomic DNA is used as a diagnostic of the 5′ ends of genes.

Conservation

One of the features of exons described in Section 2.5 was that exons of the same gene are highly conserved in different species, whereas introns and other non-coding portions vary between species. Thus the demonstration that a sequence is conserved is a good indication that it is a functional one. In order to test for conservation of a sequence, DNA from a range of species is digested by restriction enzymes and the fragments separated on an agarose gel. The DNA is then transferred to a nitrocellulose filter by Southern blotting (Box 2.2) – a **zoo blot** – and is hybridized to radiolabelled fragments from the region of interest. Cloned fragments that hybridize to the DNA from other species are likely to contain genes or parts of genes. Figure 2.34 is a zoo blot showing a particular restriction fragment from the sex chromosomes hybridized to a number of vertebrate species.

▷   On the Southern blot in Figure 2.34 the male sample of each species has two bands, whereas the female has just one, more intense band. What is the explanation for this?

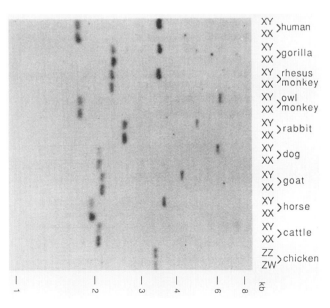

Figure 2.34 A zoo blot. Male and female DNAs from a range of species are shown. The probe is derived from a region of X–Y homology. (Note that the sex chromosome constitutions of normal male and female birds are respectively ZZ and ZW.)

▶ The probe detects a sequence which is similar but not identical, on the X and the Y chromosomes. Thus males have the X-specific band and an additional Y-specific band, and females two copies of the X-specific band.

Transcription

Of course, for genes to be expressed, they must first be transcribed. To show whether a particular fragment of DNA is transcribed, it is radiolabelled and hybridized to a **Northern blot**. A Northern blot consists of RNA transferred to a nitrocellulose filter in a similar way to a Southern blot. DNA sequences that are transcribed will hybridize to an area of the Northern blot where there are complementary RNA sequences. If a gene is being assessed as a cause of a disease, it must usually be shown to hybridize to RNA from the affected tissue(s). For example, in the search for the cystic fibrosis gene (*CF*), three conserved sequences which were suspected as being *CF* were tested on Northern blots of RNA from different tissues. Only one of these hybridized to RNA from sweat gland cells of cystic fibrosis patients (these cells are always affected in cystic fibrosis). This sequence was later confirmed as the gene affected in these patients.

2.6.2 DNA sequence analysis

It is possible to detect characteristic features of expressed sequences by looking at their nucleotide sequence. In Chapter 3 several *motifs* (short DNA sequences) will be described which adhere to a **consensus sequence** in most genes, for example the polyadenylation site, the TATA box and splice junction sequences. These short sequences may vary slightly from gene to gene but, overall, a consensus sequence can be arrived at by taking the most commonly found nucleotide at each position. All of these motifs can now be searched for in DNA sequences using computer programs. One way of identifying the exons of a gene is to look for **open reading frames** (**ORFs**). An ORF is a region of DNA which could be translated into a polypeptide without encountering a stop codon. Since three nucleotides form a codon, it is necessary to examine a particular DNA sequence starting at *each* of the first three nucleotides in turn to ensure that each of the possible reading frames have been examined. An ORF that spans around 200 nucleotides is likely to represent an exon. Computer programs to carry out this analysis are under development.

Once the exons of a gene have been identified, it is possible to infer the sequence of the protein encoded by that gene.

▷ Look at the sequence in the DNA sense strand given below in each of the three possible reading frames and determine the extent of the open reading frame. The stop codons are TAG, TAA and TGA which correspond to the RNA triplets UAG, UAA and UGA respectively. Write down the corresponding mRNA sequence and, using Table 2.3, deduce the amino acid sequence of the protein.

GATCTTGTAGTGACCATGTCTCCACTGACATTCCAGCTCTAG TTTTG

▶ The open reading frame begins with the *third* nucleotide. The first amino acid is serine encoded by bases TCT in the DNA and UCU in the mRNA respectively. The amino acid sequence is

Ser-Cys-Ser-Asp-His-Val-Ser-Thr-Asp-Ile-Pro-Ala-Leu-Val-Leu

Table 2.3   mRNA codons and corresponding amino acids.

Second letter

		U	C	A	G	
first letter	**U**	UUU UUC } Phe UUA UUG } Leu	UCU UCC UCA UCG } Ser	UAU UAC } Tyr UAA stop UAG stop	UGU UGC } Cys UGA stop UGG Trp	U C A G
	C	CUU CUC CUA CUG } Leu	CCU CCC CCA CCG } Pro	CAU CAC } His CAA CAG } Gln	CGU CGC CGA CGG } Arg	U C A G
	A	AUU AUC AUA } Ileu AUG Met	ACU ACC ACA ACG } Thr	AAU AAC } Asn AAA AAG } Lys	AGU AGC } Ser AGA AGG } Arg	U C A G
	G	GUU GUC GUA GUG } Val	GCU GCC GCA GCG } Ala	GAU GAC } Asp GAA GAG } Glu	GGU GGC GGA GGG } Gly	U C A G

third letter

The abbreviated names of amino acids are as follows: Ala = alanine, Arg = arginine, Asn = asparagine, Asp = aspartic acid, Cys = cysteine, Gln = glutamine, Glu = glutamic acid, Gly = glycine, His = histidine, Ileu = isoleucine, Leu = leucine, Lys = lysine, Met = methionine, Phe = phenylalanine, Pro = proline, Ser = serine, Thr = threonine, Trp = tryptophan, Tyr = tyrosine, Val = valine.

Knowledge of the amino acid sequence does not mean that the function of the protein will be obvious. In some cases it is possible to compare the new protein sequence to that of previously described proteins and hence be fairly sure of a function. For example, the protein structure implied by the DNA sequence of the cystic fibrosis gene was characteristic of a transmembrane protein, possibly part of an ion transport channel (see Book 3, Figure 3.13). In this way, **functional domains** that are common to a number of proteins can be identified (Book 1, Chapter 3). Other functional domains include DNA binding regions, active sites of enzymes and ligand attachment sites on receptor molecules.

2.6.3   The Human Genome Project

It is the aim of the Human Genome Project to determine the complete nucleotide sequence of the three billion (3×10^9) nucleotides of the human genome. It is estimated that this will be achieved over a period of 15 years from the autumn of 1990. The cost of the enterprise is estimated to be in the region of $3 billion, a dollar per base, but may decrease as advances in technology allow the work to become faster and more automated.

This ambitious goal will be achieved by the collaboration of scientists from all over the world. Many separate groups will clone and sequence DNA from a variety of human sources. Thus, the final sequence will not represent the DNA of one individual but a combination of many individuals. It will be a project for the future to determine which and how many nucleotides differ between different individuals, representing natural polymorphism.

The project is being coordinated in the United States by a joint committee of representatives from the National Institutes of Health and the Department of Energy. In the United Kingdom the Medical Research Council has made substantial funds available for this project.

The original sequencing project was criticized on the grounds that much effort would be expended in sequencing apparently meaningless lengths of intron and other non-transcribed sequences. Currently, the programme is focused on the construction of contig maps (Figure 2.21c) of all 24 chromosomes (22 autosomes plus the X and Y chromosomes, which are different). These contig maps are being constructed using predominantly YAC and cosmid clones (as described in Section 2.3.5). Recently, a group working in France at the Genethon laboratory has published a map of virtually the whole genome in YAC clones. At present it is not possible to sequence directly from YAC clones, and so these are being further subcloned prior to sequencing. Other maps are still being constructed. Much of the work involved at present is in the isolation and localization of new genetic markers. This is achieved by isolating polymorphic markers such as CA repeats and localizing them on the genetic map or by localizing other markers such as sequence tagged sites (Section 2.4.1) on physical maps.

The aim is to use published STSs for the whole genome so that different groups can compare their data on the basis of common landmarks. One of the goals of the project for the first five years is to isolate and localize STSs for each of the markers on the genetic map with markers spaced at 2–5 cM ($\sim$2–5 $\times$ 10^6 bp), and to assemble physical maps of all chromosomes with markers spaced at 10^5 bp intervals.

In terms of sequencing, the current goals are to develop methods that will allow large-scale sequencing at a cost of around 50 cents per base pair (that is, to halve the current cost) and to determine the sequence of large (10^7 bp) continuous stretches of DNA in parallel with developing the technology.

Pilot projects for some of the mapping techniques are being carried out using several so-called 'model' organisms, such as the nematode worm *Caenorhabditis elegans* and the yeast *Saccharomyces cerevisiae* (these have very small genomes and so are already the subject of intense genetic research), as well as the laboratory mouse. Such projects also serve as a means of interpreting data obtained in humans and understanding the human biology. It is important not to forget that the project does not end when all of the DNA is sequenced. Then begins the problem of identifying expressed sequences, as discussed above, and understanding what function these genes have and how they are regulated and interact with each other.

Objectives for Chapter 2

After completing this chapter you should be able to:

2.1   Define and use, or recognize definitions and applications of, each of the terms printed in **bold** in the text.

2.2   Have some knowledge of the structure of chromosomes, what a human karyotype looks like and how abnormalities in chromosome number can lead to disease.

2.3   Appreciate the difference between genetic and physical mapping, but understand how the two approaches are used in conjunction to isolate genes.

2.4   Explain the requirement for polymorphism in genetic mapping and describe some types of polymorphic sequences.

2.5   Describe how sequences can be assigned to a chromosome using fluorescence *in situ* hybridization and somatic cell hybrids.

2.6   Understand the principles of DNA cloning and the assembly of cloned DNA fragments into overlapping sets.

2.7   Discuss what proportion of the DNA is protein coding; the types of sequence found in the rest of the DNA, and what function, if any, these sequences might have.

2.8   Suggest ways of finding expressed sequences in a cloned region of DNA.

2.9   Describe the goals of the Human Genome Project.

Questions for Chapter 2

Question 2.1   (Objectives 2.1 and 2.2)

Describe some examples of human aneuploidy. What is the phenotype of the person who has the karyotype shown in Figure 2.35?

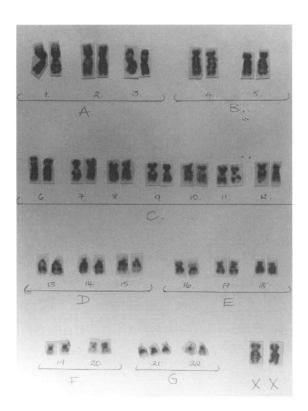

Figure 2.35   Karyotype of an individual with trisomy of chromosome 21.

Question 2.2 *(Objectives 2.3–2.6 and 2.8)*

How would you go about identifying a gene that is responsible for a disease condition:

(a) if you know which enzyme is missing in the patients and can test for its activity but do not know on which chromosome the corresponding gene is found?

(b) if you have a series of large families containing affected individuals, but no knowledge of which protein may be involved?

(c) if you have cloned a region of DNA that is deleted in all of the affected individuals?

Question 2.3 *(Objectives 2.1 and 2.6)*

What is a contig? How can a contig be of use in gene cloning?

Question 2.4 *(Objective 2.7)*

'The majority of eukaryote DNA is junk.' Discuss this statement.

Question 2.5 *(Objective 2.9)*

The following questions concern the Human Genome Project.

(a) What benefit will be gained from the results of the project?

(b) Is the project worth the money invested in it?

(c) Is it an ethically sound ambition?

Gene control

Each nucleated cell of an organism has a complete complement of genetic material. In other words, every cell contains the same DNA as every other cell. Why, then, aren't eukaryotic organisms just a mass of many cells of the same type? Humans have between 150 and 200 highly specialized tissues and each is distinct from the others. However, cells within a tissue are limited to a small number of different types. This specialization is the result of **differentiation**, which is, in turn, the result of selective gene expression giving rise to specific cell types. So, although every cell has the potential to express all of the genes in the genome, only a certain proportion of the genes are expressed in any one cell type. Furthermore, different genes may be expressed at different times. Such *spatial* and *temporal* regulation of gene expression is an extremely important feature of the genetics of a multicellular organism.

3.1 From gene to protein

In all organisms the majority of information coded in DNA is converted to protein molecules (however, you will recall that some genes encode functional RNA molecules – tRNAs and rRNAs). Both development and the essential metabolic processes of life depend on the action of proteins. As you know, the double-stranded DNA molecule consists of strings of **nucleotides.** At the simplest level, the conversion of the genetic code into a particular protein (**expression**) involves two processes, **transcription** and **translation**. We begin by recapitulating the story of protein synthesis, much of which is already familiar to you from the Science Foundation Course and Chapter 5 in Book 1 of this course.

3.1.1 Transcription

Transcription produces an RNA copy of the information held in the gene. RNA (ribonucleic acid) differs from DNA (*deoxy*ribonucleic acid) in that the ribose sugars of the nucleotides retain their 2′ hydroxyl group. In DNA there is only a hydrogen atom at this position (see *Source Book*). Thymine (T) is not found in RNA, but is replaced by uracil (U) which has a very similar structure but lacks a methyl group. The process of transcription produces a single-stranded copy of one strand of the DNA double helix (the RNA transcript). The DNA strand that has the same sequence as the mRNA (apart from possessing T instead of U) is called the **coding** or **sense strand**. The other, the **antisense strand**, is used as the template for RNA production (Figure 3.1).

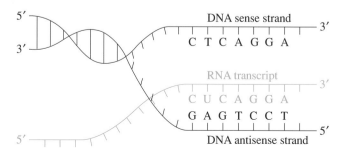

Figure 3.1 Transcription. The RNA transcript has the same sequence as the sense strand except for the substitution of U for T.

As you learnt in Chapter 2, in eukaryotes the primary transcript of a gene contains sequences (introns) which must be removed before the mRNA is transported to the cytosol. Prokaryotes do not have a nucleus, their transcripts do not contain introns and, unlike eukaryotes, they may produce a *polycistronic* mRNA molecule containing the information for several proteins.

Transcription is catalysed by an enzyme called **RNA polymerase** which incorporates ribonucleotides into a chain using the antisense strand as a template and following the usual rules of base pairing. In prokaryotes there is just one RNA polymerase, but in eukaryotes there are several versions responsible for transcribing different classes of gene. RNA polymerase II (pol II) transcribes genes that code for proteins (together with some small nuclear RNAs that are involved in the removal of introns from the primary transcript – see later). The two others, which we will be less concerned with, transcribe genes that produce an RNA, rather than a protein, final product. RNA polymerase I (pol I) transcribes the ribosomal RNA genes (except the 5S ribosomal RNA) and RNA polymerase III (pol III) transcribes genes coding for the transfer RNAs (tRNAs) and 5S rRNA.

Several stages of **processing** of the eukaryotic primary transcript occur. These will be discussed in more detail later in this chapter. They involve the removal of the intervening or intron sequences (splicing), and the addition of a 5′ cap (a methylated GTP) and a 3′ poly A tail (a string of adenosines) to form mature messenger RNA (mRNA). This is transported from the nucleus to the cytosol where *translation* takes place (Figure 3.2).

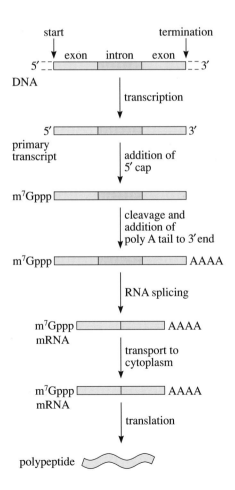

Figure 3.2 Stages in eukaryotic gene expression.

3.1.2 Translation

The process of **translation** – the production of a protein molecule using mRNA as a template – can be divided into three stages: initiation, elongation and termination. The process of translation itself is carried out at the **ribosome**. These macromolecular complexes are made up of **ribosomal RNAs (rRNAs)** in combination with a number of **ribosomal proteins**; together they make up two subunits (the *large* and *small* subunits) which clamp onto the mRNA molecule. In eukaryotes the first feature of the mRNA to be recognized by the ribosome is the 5′ cap. The ribosome then migrates downstream until it reaches a specific AUG codon near the 5′ end of the mRNA. This is the point at which translation begins and always specifies the amino acid methionine at the amino- or N-terminus of the protein.

The genetic code held in the mRNA is read three letters at a time – each group of three nucleotides specifies one particular amino acid. Such a triplet of nucleotides is termed a **codon**. Several different codons may specify the same amino acid (Table 2.3 in Chapter 2). This is because the nucleotide at the third position is not crucial, thus CUU, CUC, CUA and CUG all specify the amino acid *leucine*; this reduced specificity is known as **third base degeneracy** or 'wobble'.

Amino acids are brought to the ribosome by **transfer RNA (tRNA)** molecules. These hold an amino acid at one end and at the other there are three nucleotides that can pair with an mRNA codon. The triplet of nucleotides on the tRNA is called an **anticodon**. tRNA molecules carrying different amino acids have different anticodons. Anticodons are responsible for making sure that the correct amino acid goes in the right place in the polypeptide chain. For example, the tRNA molecule carrying leucine has a GAG anticodon which pairs with the CUC triplet in the mRNA (Figure 3.3).

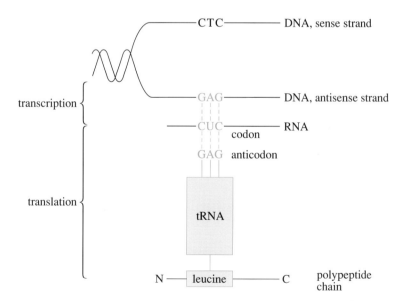

Figure 3.3   The triplet CTC in the DNA sense strand specifies the incorporation of leucine into the peptide via the CUC triplet in the mRNA which is recognized by the GAG anticodon of the tRNA molecule.

▷   Look back at Table 2.3 in Chapter 2 to confirm that the mRNA codon CUC specifies the amino acid leucine. What would be the anticodon of a tRNA molecule carrying alanine?

▶   There are four possibilities for the anticodon; GCU, GCC, GCA or GCG. This is another example of third base 'wobble'.

The initiation of translation involves the cooperation of a number of accessory proteins, termed **initiation factors** (IFs in prokaryotes, eIFs in eukaryotes) which catalyse binding of the initiator tRNA molecule carrying the first methionine to the small (40S) ribosomal subunit. The discussion that follows relates to protein synthesis in eukaryotes. First GTP binds to a factor (eIF-2) causing it to have increased binding affinity for the tRNA molecule carrying methionine (Met-tRNA) and then this ternary complex associates with a free 40S ribosomal

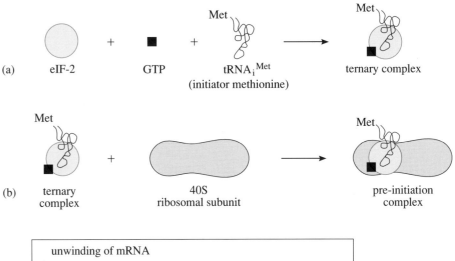

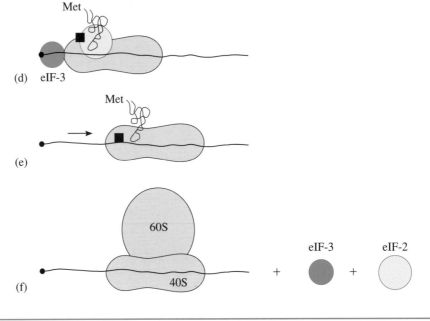

Figure 3.4 Initiation of translation. (a) GTP binds to eIF-2 to increase its binding affinity for the initiator methionine, tRNA$_i$^{Met}. (b) The ternary complex associates with free 40S ribosomal subunits, to form a pre-initiation complex. This complex binds with mRNA from which the secondary structure has been removed by unwinding. (c) The unwinding of mRNA: this is achieved by the successive binding of CAP binding protein and factors eIF-4A and eIF-4B. Factor eIF-3 is also involved. (d) eIF-3 is required for the 40S subunit with the ternary complex (the pre-initiation complex) to bind to the 5' end of mRNA. (Factors eIF-4A and eIF-4B are still involved, but are not shown.) (e) The 40S subunit migrates along the mRNA to the AUG codon. (Factors eIF-2 and eIF-3 are still here, but are not shown.) (f) Factor eIF-5 (not shown) releases factors eIF-2 and eIF-3 and allows the 60S subunit to join.

subunit. Several additional factors are required for this complex to bind to mRNA at the 5′ cap site (some way upstream from the translation starting point). This *pre-initiation complex* then moves along until it reaches the first AUG codon where translation is due to begin. The pre-initiation complex then binds with the 60S subunit in a reaction which requires energy and involves more proteins acting as accessory factors and more energy to create the functional ribosome. (Figure 3.4 shows the details of the initiation process – you are not required to remember these.)

The amino acids, delivered by the tRNA molecules in the correct order to the ribosome, are linked together via peptide bonds to make a growing peptide chain (Figure 3.5). Like initiation, the elongation process involves GTP, as well as a number of **elongation factors** (eEFs).

Termination takes place at any of the three 'stop' codons (UAG, UAA and UAG) and again involves GTP as well as, in eukaryotes, a single **release factor** (eRF). Release involves getting the protein away from the last tRNA molecule and also getting the ribosome off the mRNA.

At this stage the process may still not be mature – it may be subjected to *post-translational processing*: segments may be cleaved away by proteolysis or the protein may be modified by acetylation or glycosylation.

The conversion of genotype to phenotype – that is, DNA to protein – involves a lengthy series of steps. Control of gene expression may occur at any one, or more, of the steps shown in Figure 3.2. This allows finely tuned and complex regulation. The control of eukaryotic gene expression will be examined later in this chapter, but first we will look at the simpler control mechanisms that operate in bacterial cells. As we have said, prokaryotic genes do not have introns and in several other respects, prokaryotic and eukaryotic transcripts are processed differently. Nevertheless, many features of prokaryotic gene regulation are shared with eukaryotes.

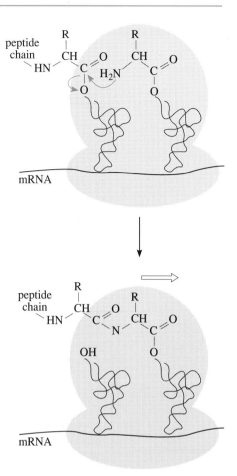

Figure 3.5 Illustration of peptide bond formation between the growing peptide attached to a tRNA molecule (peptidyl tRNA) and the amino acid of aminoacyl tRNA. The result is a free tRNA molecule and a peptide chain (attached to a tRNA molecule) which is one amino acid longer. As the ribosome moves along the RNA the next amino acid is brought alongside. (The open arrow denotes the direction of translation.)

3.2 Prokaryotic gene expression – the operon

As you will recall from Book 3, Chapter 1, bacterial cells respond to the challenges of their environment by altering the repertoire of proteins they produce. This allows efficient use of their energy and resources. A classic example of bacterial gene regulation is the enzyme system responsible for metabolism of lactose in *E. coli*. Lactose is a disaccharide which, to be metabolized, must first be split into its two constituent monosaccharides, glucose and galactose. This hydrolysis is carried out by the enzyme β-**galactosidase**. Two additional enzymes are involved in the overall process: **permease** – as the name implies – facilitates the entry of lactose into the cell and, although it does not appear to be essential, **transacetylase** (which is involved in transfer of an acetyl group from acetyl CoA to β-galactosides) may be helpful under certain conditions.

The genes encoding these enzymes are located one after another in the *E. coli* genome and are regulated together. They form an **operon**. When lactose is absent, the gene products are produced at low levels but when grown in a lactose-containing medium enzyme production is rapidly increased. Indeed, in the case of β-galactosidase, there may be 5 000 enzyme molecules per cell. When lactose is removed, enzyme production is reduced again. Such control is called **induction**, and the lactose molecule is an **inducer**. Figure 3.6 shows the arrangement of the three genes and of the DNA sequences that control their expression. The three genes encoding the enzymes involved in lactose metabolism (*lac Z*, *lac Y* and *lac A*) are all transcribed into a single mRNA molecule. This makes sense since they are all required at the same time and only then.

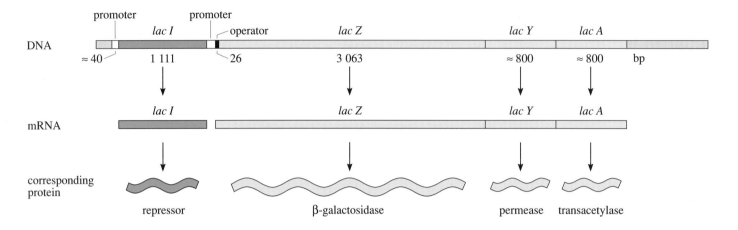

Figure 3.6 The structural and regulatory genes of the *lac* operon. The organization of these genes in the DNA is shown along with the mRNA species and proteins produced from them. Promoters of transcription are also shown, as is the operator region.

▷ How does the mRNA molecule produced from the *lac Z*, *lac Y* and *lac A* genes differ from that produced in eukaryotes?

▶ In this case, the mRNA molecule is *poly*cistronic (coding for more than one protein). In eukaryotes each mRNA is *mono*cistronic – it contains the information for a *single* polypeptide.

The genes of the *lac* operon are controlled at the level of transcription: if there is no inducer present then the genes are not transcribed. Transcription starts at the **promoter** and stops at the terminator (not shown in Figure 3.6). As this mRNA is degraded very rapidly, a cessation of transcription quickly cuts enzyme production.

The *lac I* gene at the 5′ end of the *lac* operon is transcribed from a separate promoter into a short mRNA molecule coding only for a single protein which causes transcription of the *lac Z*, *lac Y* and *lac A* genes to be stopped when lactose is not present. This molecule is a *regulator protein* which acts as a **repressor**. The genes that encode controlling proteins are called **regulatory** rather than **structural** genes.

▷ What would be the cellular effect of a mutation in a regulatory gene compared to that in a structural gene?

▶ Both *gene products* would be similarly affected – that is, there would be a lack of product, or an altered product would be made. The effect on the cell, however, would differ so that a mutation in a regulatory gene might alter the levels of transcription of *a number of* genes under its control; a mutation in a structural gene might result in a lack of, or change in, that *single* gene product.

The repressor molecule is a tetramer. It acts by binding to a region of about 26 base pairs immediately downstream of the promoter, called the **operator**. When the repressor is bound in this position it blocks binding of RNA polymerase, which then fails to initiate transcription (Figure 3.7a).

When molecules of lactose are present, the low levels of β-galactosidase present in the cell convert some molecules to allolactose. These molecules then associate with the repressor molecule and convert it to a form which cannot bind to the operator. This allows the RNA polymerase to bind and transcription to commence so that the three lactose metabolism enzymes are synthesized (Figure 3.7b).

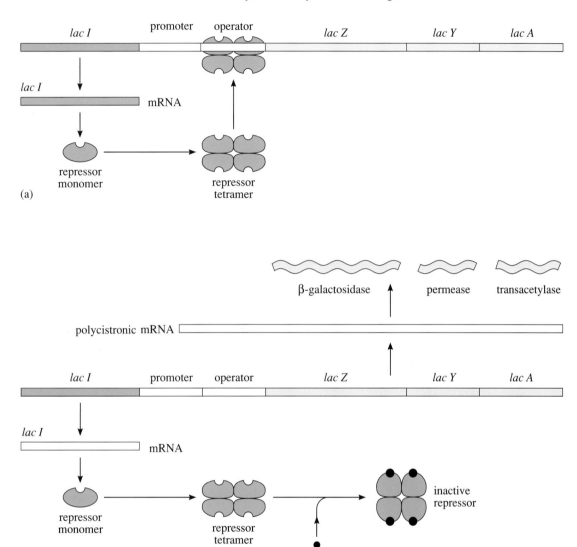

Figure 3.7 (a) In the absence of lactose, the tetrameric repressor protein binds to the operator and prevents transcription from taking place. (b) When lactose is present, this binds to the repressor (in the form of allolactose) and inactivates it – that is, it no longer binds to the operator. When the repressor is not bound to the DNA at the operator the transcription machinery can go ahead to produce the polycistronic mRNA transcript which is translated into the three enzymes shown.

▷   What would happen to expression levels if the *lac I* gene were mutated so that no product could be formed?

▶   There would be continuous or **constitutive** production of the three *lac* genes. Constitutive mutations were in fact important in working out just how the *lac* operon functions.

▷   What other mutation would result in constitutive expression?

▶   A mutation in the operator region which prevented binding to the repressor.

The interaction between the lactose repressor and operator is a good model of a **sequence-specific DNA-binding reaction**. These are very important in both prokaryotic and eukaryotic gene expression. Many DNA sequences that are recognition sites for proteins have twofold rotational symmetry (just like the recognition sites for restriction enzymes – Box 2.1). For the *lac* operator there is a sequence 17 bases on either side of the axis within which most of the bases are symmetrical (Figure 3.8).

Figure 3.8 Symmetry in the sequence of the *lac* operator. Regions of twofold rotational symmetry are indicated by the shaded blocks. Look back at Figure 5.12 in Book 1, to recall the cruciform structures that inverted repeats can form. Notice that transcription actually starts within the operator region.

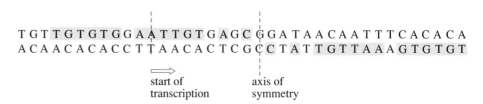

It is probable that the two halves of the tetrameric repressor protein each bind to the two halves of the recognition sequence. Mutations at different positions in the *lac I* gene suggest that the N-terminal ends of the regulatory protein subunits are involved in binding to the operator sequence. Thus, mutations at the 5′ end of *lac I* result in a protein that is unable to bind to the operator sequence. These mutations are *dominant* – one 'bad' subunit is sufficient to prevent binding even when associated with three 'good' subunits. Some rare mutations in this region have the opposite effect. These 'tight binding' mutations result in a repressor molecule which cannot be released from the operator even in the presence of the inducer. Mutations towards the 3′ end of the gene (the C-terminal ends of the repressor protein subunits), on the other hand, affect binding of allolactose to the repressor. It is these regions which must be involved in inducer binding.

Once inducer molecules are present, they bind to the repressor protein attached to the operator. Binding of just two molecules of inducer to two of the four potential binding sites (Figure 3.7) is enough to release the repressor. Release is brought about by a conformational change in the inducer binding regions which in turn affects the N-terminal DNA-binding region, so that it no longer binds tightly to the operator sequence (Figure 3.9).

Although the ability to act as a repressor or an inducer of an operon is highly specific, some molecules may activate transcription of an operon – even though they are not involved in that system – if they closely resemble the natural inducer molecule. For example, the *lac* operon can be induced by a galactose derivative called isopropylthiogalactoside (IPTG). These types of molecules which are not metabolized themselves are called **gratuitous inducers**.

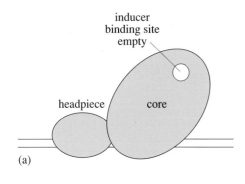

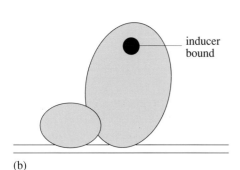

Figure 3.9 (a) Diagram showing the two regions of a repressor molecule. The 'headpiece' is responsible for DNA binding, the 'core' for binding of inducer. (b) When an inducer molecule (black circle) is associated with the core a conformation change occurs with the result that the headpiece can no longer bind tightly to the DNA and is released.

In summary then, the structural genes involved in lactose metabolism are jointly regulated by induction of the operon by the metabolite. The amounts of the three enzymes present in the cell at any one time are controlled by the rate of transcription. This in turn is regulated by a molecule (the repressor) which binds to a specific DNA sequence (the operator) only in the absence of inducer.

Several similar control systems exist in bacteria. Some, like the *lac* operon, are said to be under **positive control**; the genes are not expressed unless an active regulator protein is present. Some operons are under **negative control** – the genes are expressed until they are switched off by a repressor protein (which may be the end product of a metabolic pathway – see Book 3, Chapter 1).

▷ The tryptophan (*trp*) operon includes five genes required for the conversion of chorismic acid to tryptophan. An *inactive* repressor protein is produced which, in the presence of tryptophan, is activated to bind the operator DNA and inhibit transcription. Is this operon under positive or negative control?

▶ When tryptophan is not present the genes are expressed and the presence of tryptophan halts expression. Therefore, this is an example of negative control.

3.3 Eukaryotic gene expression

Eukaryotic genes are divided into three parts : the **5′ flanking, upstream** or promoter region which precedes the start site of transcription and largely controls its initiation; the transcribed region which includes several open reading frames (the exons) and specifies the protein; and a **3′ flanking, downstream** region which controls the termination of transcription (Figure 3.10).

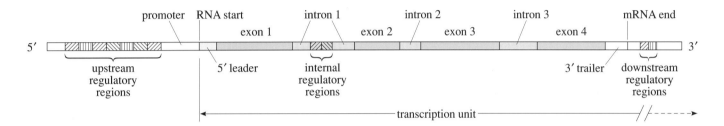

Figure 3.10 Different regions of a eukaryotic gene. Notice the regulatory regions, which may be in the 5′ upstream region, the 3′ downstream region or within the introns of the gene itself. Transcription starts usually just downstream of the promoter, thus excluding the upstream regulatory regions but may continue into the 3′ downstream regions.

3.3.1 Transcription

As in prokaryotes, the sequence upstream of the transcription start site in eukaryotes is called the promoter and is instrumental in the control of gene expression. Promoter action is mediated by certain proteins called **transcription factors** which bind specifically to *sequence motifs* – short stretches of DNA (~10 bp) in the promoter. The sequence motifs act in concert with each other and with the transcription factors to control: the site where transcription begins; the rate of transcription initiation and the spatial and temporal pattern of transcription in response to environmental or developmental changes or cues. These sequences are *cis* acting – they affect the genes that are physically contiguous with them. The transcription factors are *trans* acting – they are diffusible products which can act on any relevant sequence whether present on the same or different molecules of DNA.

Sequence motifs

Although the structure of the promoter varies from gene to gene, they usually contain certain sequences in common: in eukaryotes a 7 bp sequence called the **TATA box** is most common. This consists of As and Ts, and is usually found about 20–30 bp upstream of the transcription start site. Many promoters also contain **upstream motifs**, located at about 80–100 bp upstream of the start site, that determine the efficiency of transcription. Some are common to many genes. The **CCAAT box** and the **GC box**, for example, may be found individually or together and sometimes in multiple copies (Figure 3.11). Some upstream sequence motifs are promoter-specific or are restricted to certain gene families. For example, when an organism (eukaryotic or prokaryotic) is subjected to thermal or chemical stress, a set of genes becomes active simultaneously in what is known as the **heat shock response**. The proteins produced from these genes presumably play an active role in protecting the cell from the deleterious effects of these stresses. The promoter regions of heat shock protein genes contain **heat shock response elements (HSE)** in the section within about 200 bp upstream of the TATA box which regulate the response in a coordinated way. The consensus sequence (that which represents the most commonly found nucleotide at each position) of the HSEs is 11 nucleotides long and, like other regulatory elements, has twofold rotational symmetry (5'-CNGAANTTCNG-3', where N can be any base).

Regulatory proteins bind to the HSEs in response to a heat shock and hence activate all genes containing heat-inducible promoters. This occurs in preference to transcription from other genes from which transcription is reduced.

Some heat shock genes are also expressed constitutively at lower levels, when their expression is under the control of additional regulatory sequences in their promoters. Thus eukaryotic genes may have a number of different regulatory sequences in their 5' region all responding to different 'stimuli' (Figure 3.11): a system considerably more complex than the *lac* operon.

Figure 3.11 Transcriptional control elements in the *hsp 70* gene promoter. The names of the sequences are below the line and the transcription factors that bind to them are above. Notice the heat shock response element (HSE) to which the heat shock factor binds (HSF) but also the number of other control elements some of which, such as the TATA box, are found in very many genes. (The numbers indicate the number of base pairs upstream of the transcription start site at which each element may be found.)

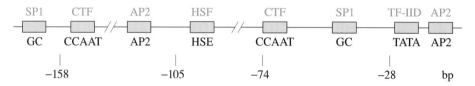

▷ How does the coordinated regulation of the enzyme genes of the *lac* operon differ from that of the heat shock genes?

▶ The genes of the *lac* operon share one promoter and are transcribed as one unit. The heat shock genes are scattered around the genome, each with their own promoter. Coordination occurs because the promoters have sequences in common and all respond to the same inducers.

Enhancers are sequence motifs that act as activators of transcription, (although, in some circumstances, they can repress it). They are usually some distance (often several kilobase pairs) from the transcription start site, either upstream or downstream, but also exist in the transcribed region and 3' flanking region. Enhancers are tissue- rather than gene-specific – if the 72 bp enhancer sequence of the SV40 virus is linked to the β-globin gene, transcription of β-globin occurs in tissues where it normally would not. Enhancers associated with some genes are active only at certain developmental stages and regulate transcription by being active only under certain physiological conditions. Enhancers function in the same way as other *cis* acting regulatory sequences by binding transcription factors.

Some sequences behave like enhancers in that they are independent of orientation and distance but always *repress* transcription. These are known as **silencers**.

3.3.2 Transcription factors

Promoters, enhancers and silencers work by binding to proteins called transcription factors. We have already seen how proteins binding to the operator sequences of the *lac* operon can have positive or negative effects on transcription. In eukaryotes, a number of transcription factors are required for transcription to start. These are assembled in the **pol II transcription complex** (Figure 3.12). These include a protein called TF-IID (the TATA box binding factor – Figure 3.11) which binds to the TATA box. This basal transcription complex allows only a very low level of transcription to occur, if any.

When an additional factor binds to its own specific sequence element (for example the GAL4 transcription factor of yeast which responds to galactose), binding of TF-IID alters so that it also encompasses the transcription start site. The binding of TF-IID at this point enables the other transcription factors as well as the RNA polymerase to form a stable complex which then proceeds to transcribe the gene at a much higher level.

Transcription factors that bind to promoter elements a long way from the start site must be able to influence the activity of RNA polymerase at a distance. This probably involves the looping of DNA so that the transcription factors come into contact with each other and with the polymerase (Figure 3.13).

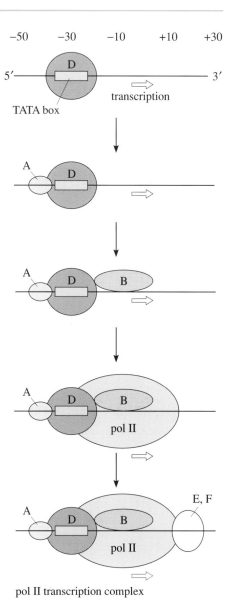

Figure 3.12 The role of transcription factors in the assembly of a transcription complex at the site of transcription initiation. The complex includes factors TF-IIA, B, D, E and F and RNA polymerase II (pol II). The transcription start site is indicated by the open arrow. (Numbers denote distance in base pairs upstream or downstream of the transcription start site.)

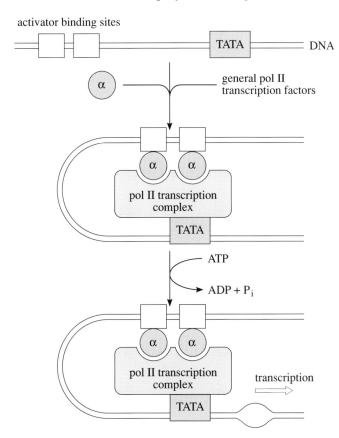

Figure 3.13 Activators (α) bind to their recognition sites some distance from the TATA box. If the DNA loops around, these can come into contact with the other components of the pol II transcription complex (i.e. pol II plus transcription factors). It is likely that multiple upstream activators simultaneously interact with different members of the general transcription machinery. When all of these interactions have taken place and energy has been released (via ATP hydrolysis), transcription can take place.

The central feature of transcription factors is that they bind DNA in a sequence-specific manner, and influence transcription by interacting with other transcription factors or with the RNA polymerase itself. Transcription factors need at least two different protein domains: the DNA binding domain and the activation domain which interacts with the RNA polymerase transcription complex and activates transcription.

3.3.3 Different types of transcription factor

Different transcription factors bind to different consensus sequences. The transcription factor needs to recognize a specific base sequence and bind to the DNA double helix. Different transcription factors have different **DNA binding domains** which enable them to do so. Several factors may share similar domains and they are often classified according to their structure. DNA binding domains include the *helix-turn-helix*, the *zinc finger* and the *leucine zipper* motifs. These were introduced in Book 1, Chapter 5, and examples of each are given below.

Helix-turn-helix

There are many developmental mutations in *Drosophila*. Sometimes they lead to the **ectopic** (misplaced) expression of a particular group of genes, the *homeotic* genes. In one, the mutant fly has legs in place of antennae.

▷ In which type of gene must these mutations lie?

▶ In genes responsible for determining the overall body plan of the organism.

The early embryo becomes divided into a series of segments and the homeotic genes act to assign an identity to each of these regions. When their normal function is disrupted, the identity assigned to each segment is not correct for its location.

The homeotic gene products act at particular times in development to activate or repress the expression of other genes, binding to regulatory regions in a sequence-specific manner. Such genes are found in a wide range of organisms besides the fruit-fly, including amphibians, birds and mammals. Homeotic genes possess a common sequence of about 180 bp encoding 60 amino acids called the **homeobox** or **homeodomain**, which is highly *conserved* (unchanged) through evolution so that the sequences are very similar in mice and humans. The three-dimensional structure of this peptide is an α-helix followed by a β-sheet and then another α-helix, a **helix-turn-helix** motif (Figure 3.14a). When the homeodomain binds DNA, one helix lies across the DNA major groove and the second lies partly within the major groove where it makes specific contacts with the bases (Figure 3.14b). This second helix – the **recognition helix** – is responsible for sequence-specific DNA binding. Such structures are also found in other regulatory proteins, including those found in bacteria.

Zinc fingers

The TF-IIIA transcription factor, which is involved in transcription of the 5S RNA, has a series of nine finger-like protrusions each of which has a zinc atom held between four residues, two cysteine and two histidine. For obvious reasons this DNA binding domain is called a **zinc finger** motif.

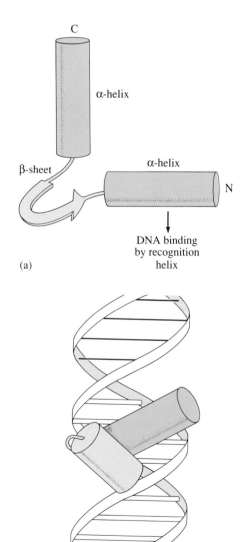

Figure 3.14 (a) Structure of the helix-turn-helix motif found in some transcription factors. (b) The helix-turn-helix motif binding to DNA.

Glucocorticoids are steroid hormones that promote *gluconeogenesis* (synthesis of glucose from non-carbohydrate precursors) and the formation of glycogen, and enhance the degradation of fat and protein. *Oestrogens* are the steroid hormones responsible for the development of female secondary sexual characteristics. The receptors for such steroid hormones belong to a family of proteins with a highly conserved central domain which is responsible for DNA binding. The C-terminal region binds the appropriate hormone, and the N-terminal region activates transcription of the target genes (Figure 3.15). The DNA binding domain of these receptors resembles the zinc fingers described above, but this time the zinc atom is held by four cysteines – the cysteine zinc finger. These receptors contain two fingers only. The first finger is responsible for the specificity of DNA binding (Figure 3.16).

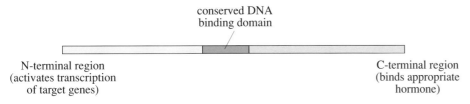

conserved DNA
binding domain

N-terminal region
(activates transcription
of target genes)

C-terminal region
(binds appropriate
hormone)

Figure 3.15 Domain structure of the members of the steroid/thyroid hormone receptor family. This is a family of related genes encoding hormone receptors (e.g. steroid hormones, thyroid hormones, retinoic acid).

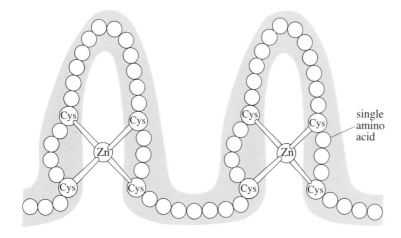

single
amino
acid

Figure 3.16 Structure of the four-cysteine zinc finger.

Two receptor molecules bind to the DNA recognition sequence, which has twofold rotational symmetry. The oestrogen receptor recognizes almost the same DNA sequence as does the thyroid hormone receptor; except that, in the former, the two halves are separated by three nucleotides (Figure 3.17).

thyroid hormone
sequence   --T C A G G T C A T G A C C T G A--

oestrogen sequence -- A G G T C A N N N T G A C C T --

Figure 3.17 Consensus sequences of the oestrogen and thyroid hormone-responsive elements. Notice that the sequences show twofold rotational symmetry. In the oestrogen receptor the two halves of the sequence are separated by three nucleotides.

If the amino acid sequence of the second finger of the oestrogen receptor is altered, it responds to thyroid hormone. This is presumably because the second finger is responsible for spacing the two receptor molecules and when this is altered the protein dimer can bind to the two closer halves of the thyroid receptor recognition sequence.

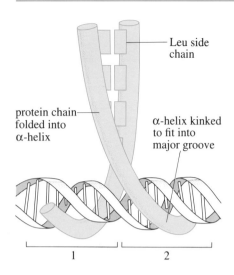

protein chain folded into α-helix

Leu side chain

α-helix kinked to fit into major groove

1

2

Figure 3.18 The leucine zipper. Leucine residues interact to bring about dimerization of two leucine zippers. The basic regions of the proteins can then bind to the DNA.

Leucine zippers

A **leucine zipper** is characterized by stretches of amino acids with periodic leucines which, when the protein forms an α-helix, occur every two turns on the same side of the helix. Two such proteins can form a dimer with the leucines interdigitating with one another (Figure 3.18). Once two molecules of a transcription factor have dimerized in this way, the adjacent region of the protein (which is basic) can bind to the acidic phosphate groups of DNA in a sequence-specific manner.

Transcription factors having this structure include certain *proto-oncogene* proteins which were identified by their ability to transform cells to the cancerous phenotype (see Chapter 7). They act by regulating the transcription of other cellular genes.

3.3.4 Transcription factors – a brief review

Some general transcription factors, like TF-IID, are present in all cell types and are essential for the transcription of many genes. Other transcription factors are activated by external stimuli such as heat shocks, metabolites or hormones and then activate genes containing their binding sites. As some genes are expressed only in one tissue, they have recognition sites for transcription factors found only in that tissue. For a gene to be expressed in more than one tissue it may respond to a general transcription factor or may have different promoter elements responding to different factors. These transcription factors are, of course, themselves the result of controlled gene expression. The complexity of control is formidable, involving a hierarchy of interacting genes.

Sometimes mutations in transcription factors lead to inappropriate gene expression. We have discussed one example of ectopic expression in *Drosophila*. It is also possible for genes to be expressed at the wrong time – **ecchronic expression**. This happens when a transcription factor is altered so that it recognizes the wrong stimulus. You have seen how, *in vitro*, a difference of a few nucleotides in the zinc fingers can make the oestrogen receptor responsive to thyroid hormone.

3.3.5 DNA methylation and transcription

In Chapter 2 we saw that 40–60% of genes have a large number of the dinucleotide CpG (CpG islands) at their 5′ ends and the cytosine residues therein are often methylated. This methylation is very important in gene control. Genes expressed at all times in all tissues – the so-called housekeeping genes – have unmethylated CpG islands while genes expressed in only one tissue don't always have CpG islands, but when they do, they have islands that are unmethylated only in that tissue. This strongly suggests that methylation of the 5′ region of a gene is involved in regulating transcription. There is indeed a large protein which binds strongly to methylated CpG islands and perhaps blocks access of transcription factors and RNA polymerase to such regions.

3.3.6 X chromosome inactivation

So far, only the transcriptional regulation of single genes or of small groups of genes involved in one process (such as heat shock genes) has been discussed. Gene regulation is much more wide-ranging than this. In mammalian female somatic cells one copy of the X chromosome is transcriptionally inactive – **X**

inactivation. It forms a small condensed blob known as a Barr body. Females (XX) have one Barr body per cell, males (XY) have none; thus there is one copy of the X chromosome active in each sex. X chromosome inactivation is a device by which the levels of expression of genes on the X chromosome can be matched in males and females (of course, males have only one X chromosome). This process is called **dosage compensation**. Inactivation occurs early in embryonic development. In a female embryo it is equally likely that the X chromosome inherited from the mother or from the father will be inactivated. It means that females are effectively **mosaics** (they have cell populations with different active genetic components), at least for genes on their X chromosome. The classic example is the tortoise-shell cat. Such cats are heterozygous at an X-linked locus for a dominant allele (*O*) leading to orange hairs, and a wild-type recessive (*o*) leading to black hairs (when homozygous). Whenever a patch of cells has arisen during development from a cell in which the chromosome containing the *O* allele is *inactivated*, the fur will be black. Other patches will be orange (Plate 3.1).

▷ As can be seen from Plate 3.1, there are more than two colours to a tortoise-shell cat – why is this?

▶ There are many additional autosomal genes which collectively affect the colour and pattern of a cat's fur.

How X chromosome inactivation occurs is not yet known. At least three steps are involved; initiation, spreading and maintenance. Inactivation starts from a region on the long arm of the X chromosome (the inactivation centre), spreads along the chromosome and is maintained throughout subsequent cell generations. A gene, *XIST* (X-inactive specific transcript), is responsible for initiation but not maintenance of X inactivation. The gene is expressed only from the inactive X chromosome. Its product is an RNA molecule that is not translated. How *XIST* controls X inactivation has yet to be determined. DNA methylation is probably an important factor in the maintenance of the inactive state. The promoters of genes on the inactive X chromosome are heavily methylated, while the promoters of the same genes on the *active* X are not methylated.

There is an additional – and as yet baffling – complication. In individuals with anomalies of the sex chromosomes (for example XXY), one X chromosome is inactivated. In individuals with three X chromosomes, two of the X chromosomes are inactivated. There appears to be a 'counting mechanism' of some kind which inactivates all X chromosomes until only one is left active. This is presumably the reason why sex chromosome aneuploidies are so well tolerated (Chapter 2).

Not all X-linked genes are inactivated. Some genes on the X chromosome have a matching copy on the Y chromosome so that, in males, two copies are expressed. In females, therefore, these genes also need to be expressed in two copies, and in turns out that they are – they *escape X inactivation*. The most obvious area where this is true is the pseudoautosomal region (Chapter 2) at the tip of the X and Y chromosome short arms where the two chromosomes share all sequences. All known genes in this region of the X chromosome escape X inactivation. In addition, there are some genes buried deep in the inactivated regions that manage to escape being silenced. One, *RPS4X* (which encodes one of the ribosomal proteins), is located close to the inactivation centre itself. The promoters of the genes that escape inactivation must have some feature that either prevents them from becoming methylated, or ensures that in some other way they are still accessible to the transcription machinery.

3.3.7 Chromatin structure and gene expression

Eukaryotic DNA is always associated with a number of different proteins to form a nucleoprotein complex called chromatin (Chapter 2, Section 2.1). Two main classes of protein are involved: the histones which package the DNA into nucleosomes, and the **non-histone chromosomal (NHC) proteins**. The NHC proteins make up a diverse group with a variety of functions. Some have a structural role similar to that of histones. Some, including the **high mobility group (HMG) proteins**, are moderately abundant and are usually found in association with **active chromatin**, which is capable of being transcribed. Other NHC proteins are present in very small amounts and act as the specific transcription factors for individual genes.

Active chromatin has increased sensitivity to endonucleases (enzymes that cut the DNA up into its constituent nucleotides) in transcribed regions and for some distance upstream and downstream of the transcribed region. There are two levels of nuclease sensitivity:

Active chromatin is generally more sensitive to nuclease digestion than is inactive (non-transcribed) chromatin and gives 200 bp fragments – this reflects the condensation state of the chromatin – 'beads on a string' (Chapter 2) versus 'solenoid' structure (that is, the string of beads wound up to form a 30 nm nucleosome fibre: Book 1, Figure 5.18). Within active chromatin, the 5′ end of genes (the promoter region) may be particularly susceptible to complete degradation. Such **hypersensitive sites** often appear in particular genes just before transcription begins in that particular tissue or developmental stage. They represent regions that are free of nucleosomes and therefore, more likely to allow entry of regulatory proteins or the RNA polymerase itself. These exposed regions

Figure 3.19 (a) The hypersensitive site (marked with a cross) in the heat shock gene is present before the heat shock transcription factor binds to it, but does not allow transcription to take place until the factor (activated by heat) has bound. (b) In the case of the steroid-inducible genes it is the receptor–steroid complex which creates a hypersensitive site which can then bind an active transcription factor.

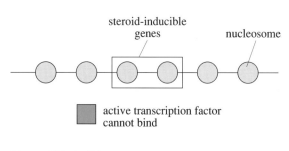

are necessary for transcription but by themselves do not allow it to take place. For example, in the genes of the heat shock response, there is a site that is hypersensitive to the endonuclease DNAase I, but the genes are transcribed only when the activated heat shock transcription factor has bound to this site (Figure 3.19a). In the case of the glucocorticoid-responsive genes, the hypersensitive site is generated by the binding of the receptor–steroid hormone complex to the recognition site, which displaces a nucleosome. The ubiquitous transcription factors (such as the TATA box binding factor, TF-IID) then bind to the region and transcription begins (Figure 3.19b).

Only about 10% of the chromatin in eukaryotes is packaged in the active 'beads on a string' conformation and is being expressed or has the potential to be expressed in a tissue. The rest is **inactive chromatin** – that is, DNA that isn't expressed in a cell type is packaged as a more dense 'solenoid' structure. In different cell types different regions of the DNA are packaged as active chromatin. This reflects the differences in gene expression in each cell type.

Some regions of chromatin are inactive in all cell types. Such heterochromatin (Chapter 2, Section 2.4.2) is normally located within the centromeric and telomeric regions of chromosomes. In mammals the inactivated X chromosome is also found to be almost entirely heterochromatin.

The change in chromatin structure associated with increased transcription can be visualized in the so-called 'lampbrush' chromosomes in oocytes (egg-producing cells) of a wide range of vertebrates. They have been studied particularly in amphibians. At a certain stage of meiosis, the chromosome is seen to consist of a series of loops which emerge from regions of condensed chromatin. The loops represent the regions of active chromatin. Indeed, on closer inspection, the loops are seen to be covered in the RNA products of transcription (Figure 3.20).

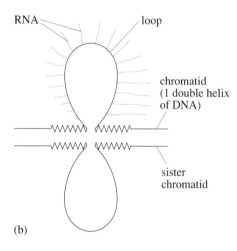

(b)

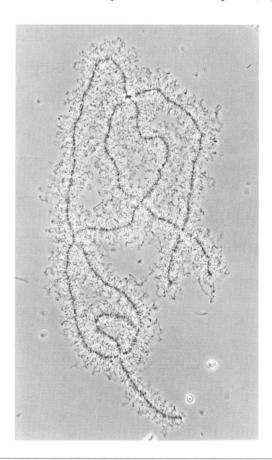

Figure 3.20 (a) Lampbrush chromosome from the newt *Triturus viridiscens*. The photograph shows the two homologous chromosomes (each of which consists of two chromatids) which are held together at various positions (the chiasmata). Notice the numerous loops of nucleoprotein extending from the central axis. (b) A diagram of part of one pair of chromatids. The loops extend out of regions of highly condensed DNA. The loops are covered in strands of RNA and are longer towards the 3′ end of the gene (right on this diagram). In this case, the strands are drawn as if a single gene is being transcribed around the whole loop – the short RNA strands are near the site of transcription initiation, the longest towards the termination point. In some cases, there are several genes per loop, and, since each loop represents one DNA helix with two strands, some genes may be transcribed from the other strand in the opposite direction.

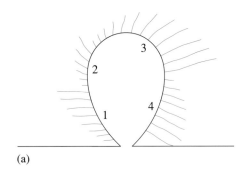

(a)

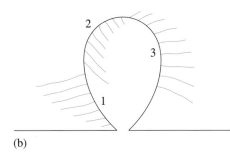

(b)

Figure 3.21 Diagrams showing some patterns of transcription from lampbrush chromosomes (see in-text question).

▷ Draw a diagram like the one in Figure 3.20b showing loops of one chromatid with (a) four genes being transcribed from one strand, and (b) three genes being transcribed from one loop with one of them transcribed from the opposite strand and in the opposite direction.

▶ See Figure 3.21.

There are many loops visible at the stage of development shown in Figure 3.20a – around 10 000 per haploid genome. On average, one-tenth of the genome is being transcribed in these cells at this time. This is a surprisingly large amount of the potentially active chromatin. Why so many genes are expressed during oocyte meiosis is not known.

3.4 Post-transcriptional processing

As you saw in Section 3.3, genes have three main parts: a 5′ upstream region, a transcribed region and a 3′ downstream region (Figure 3.10). So far we have concentrated on the upstream region and its role in the initiation of transcription.

The downstream region contains signals for the *termination* of transcription and for *post-transcriptional processing* of the 3′ end of the mRNA (removal of the terminus and addition of a poly A tail). Although it is clear that there must be specific transcription termination sites in this region of the gene, these are not easy to identify. Most pol II mRNA transcripts (those transcribed by RNA polymerase II) are **polyadenylated** – the 3′ end consists of a tail of 50–200 adenosine residues. This tail is added post-transcriptionally 10–30 bp downstream of the **polyadenylation signal**, **AATAAA**, which is involved in the processing of the 3′ end of the transcript. Most genes have another signal, the **GT element,** 30 bp downstream of the poly A site. This also plays a role in the processing of the transcript. The 3′ processing of transcripts is carried out by **small nuclear ribonuclear proteins (snRNPs).**

As you know from Chapter 2, not all the transcribed region codes for amino acids. Introns are removed from the primary transcript before it is transported to the cytosol. Even the mRNA contains portions that are not translated – a 5′ untranslated sequence or **leader** (which may be important in controlling translation) and a 3′ untranslated region or **trailer** (which may affect both translation and mRNA stability). (You may wish to refer back to Figure 3.10 here.)

The length of the leader sequence ranges between 30 and 700 bp. There is usually a post-transcriptional modification at the 5′ terminus. A methylated guanosine nucleotide, the '**cap**', is added by **guanylyl transferase**. This, too, influences mRNA stability and translational efficiency.

The trailer sequence is usually 50–200 bp long. It is important in determining mRNA stability. Those mRNAs that are expressed only transiently contain several copies of an AU-rich sequence at the 3′ ends. These mRNAs have half-lives of less than 30 minutes compared to the long-lived mRNAs such as β-globin mRNA, which have half-lives of over 16 hours and do not contain such sequences.

▷ How can mRNA stability affect the control of gene expression?

▶ Since a mRNA molecule can be translated many times, more protein will be produced if the mRNA remains in the cell for longer. Controlling the rate of mRNA degradation can thus regulate the expression pattern of the gene.

You will meet more aspects of mRNA stability in Section 3.5.3.

3.5   Control of expression at the RNA level

3.5.1   RNA splicing

RNA splicing involves the removal of the introns from primary transcript and the subsequent rejoining of the exons in the right order.

The sequences at each end of the intron are essential for splicing. The 5′ end of an intron is called the **5′ splice donor site** and almost always contains a **GU** dinucleotide. The 3′ end – the **3′ splice acceptor site** – almost always includes an **AG** dinucleotide. There is also usually a pyrimidine-rich (C+U) tract near to the 3′ splice site (this is essential in mammals). An adenosine residue near the 3′ end of introns is essential for splicing. There is no other intronic sequence requirement for splicing to occur, but the intron must be at least 80 nucleotides long.

Splicing involves two reactions. First, the phosphodiester bond between the last nucleotide of the exon and the first guanosine of the intron is broken and the guanosine is bonded to an adenosine residue towards the 3′ end of the intron. In mammalian introns any of several As within 18 to 37 nucleotides upstream of the 3′ splice site may be used. By contrast, yeast rely on a discrete sequence called the branch-point sequence located between 6 and 59 nucleotides upstream of the 3′ splice site within which is the reactive A residue (Figure 3.22). Second, the 3′ end of the first exon is joined to the 5′ end of the second exon. This releases the intron in a **lariat** (lasso) structure (Figure 3.23).

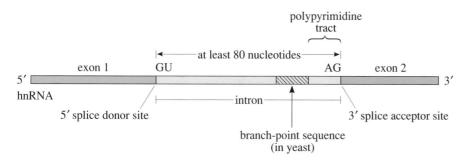

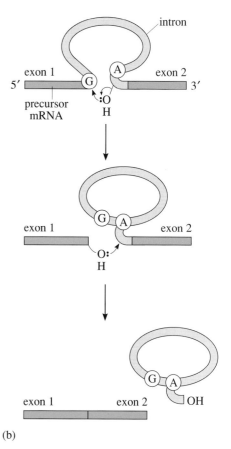

Figure 3.22   Splicing – representation of the primary transcript showing features required for accurate splicing. See text for further explanation.

Splicing occurs in the **spliceosome** – a complex of proteins and small ribonuclear proteins (snRNPs) which is formed on the primary transcript. (See Book 1, Chapter 5, Section 5.4.)

3.5.2   Alternative splicing

Most genes contain multiple introns, and each intron is processed independently; they are not removed sequentially in the 5′ to 3′ direction. There must therefore be a way of ensuring that exons are spliced together in the correct order.

This means that a particular gene transcribed in different tissues may be processed in different ways to produce different mRNA molecules (and thus different proteins). For example, a gene with four exons may be spliced to leave all its exons in the mRNA, or an exon may be spliced out along with the introns. Thus, for example, exons 1, 2 and 4 may be left or 1, 2, and 3. The alternatives may be produced in different cells or at different developmental stages.

Figure 3.23   Excision of the intron lariat. The 2′ hydroxyl group of the reactive adenosine cleaves the phosphodiester bond between exon 1 and the intron. The 3′ splice site is then cleaved by the 3′ hydroxyl group of exon 1 and the two exons are joined.

The gene encoding calcitonin (a 32 amino acid calcium regulatory protein) can also produce a different 36 amino acid protein called calcitonin-gene-related peptide (CGRP). Calcitonin is produced in the thyroid gland whereas CGRP is produced within the brain and peripheral nervous system (Figure 3.24). The occurrence of two splicing products is due to the presence of tissue-specific splicing factors. Alternative gene products are also produced when different polyadenylation signals are used in different tissues. Thus, if a polyadenylation site further upstream is used, a shorter protein is usually produced.

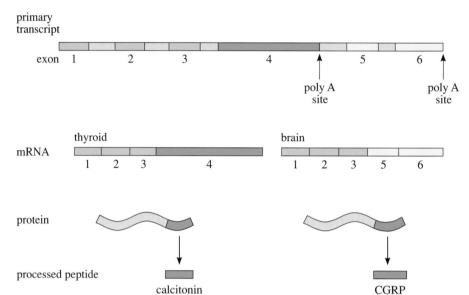

Figure 3.24 Alternative processing of the calcitonin/CGRP gene. In the thyroid gland calcitonin is produced after splicing together of exons 1, 2, 3 and 4 followed by translation and processing of the protein. In the brain exons 1, 2, 3, 5 and 6 are used. The subsequently processed peptide is CGRP.

3.5.3   mRNA stability

An example of how gene expression can be regulated by altering mRNA stability is found in the milk protein, casein. In mammary gland cells, casein mRNA has a half-life of around an hour. Following treatment with the hormone prolactin (produced by lactating mothers), the half-life of this RNA increases to over 40 hours. This results in a substantial increase of protein production merely by having the mRNA around for longer; in addition to this, the rate of transcription increases by two- to fourfold.

Sequences increasing the stability of mRNA are found in the 3′ untranslated region of many genes. The poly A tail protects the mRNA from attack by exonucleases, enzymes that 'nibble' away the ends of single-stranded nucleic acid molecules. Sequences near the 3′ end of transcripts are also important. They form **stem-loop** structures when nucleotides of the single-stranded RNA molecule pair with each other (Figure 3.25a). This structure may confer resistance to exonuclease digestion.

The protein product of the *transferrin receptor* (TfR) gene is the principal means of iron uptake into mammalian cells. When there is plenty of iron in the cells, the rate of TfR synthesis is decreased. The TfR mRNA sequence contains a region of about 680 nucleotides at the 3′ end called the **iron responsive element (IRE)**, which can form five stem-loop structures. When cells are deprived of iron, a protein (the **iron-responsive element binding protein, IRE-BP**) binds specifically to the iron-responsive element (IRE), and, in association with the stem-loop structures, prevents RNA degradation by endonuclease and exonuclease attack.

When iron is abundant, the protein does not bind and the RNA is degraded (Figure 3.25b). How does iron regulate the binding activity of the IRE-BP? The five stem-loop structures all contain a cysteine residue. To bind, the protein requires a free –SH group such as that supplied by the cysteines; in the presence of iron, S–S bridge formation between the cysteines is favoured and the IRE-BP cannot bind effectively. Hence, the level of iron regulates the stability of the TfR mRNA molecule.

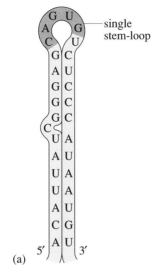

(a)

Figure 3.25 (a) A stem-loop structure. Five stem-loop structures like the one shown here occur in tandem in the 3′ untranslated region of the human transferrin receptor (TfR) mRNA. (b) Stability of transferrin receptor (TfR) mRNA. In the presence of iron, the iron-responsive element binding protein (IRE-BP) does not bind to iron-responsive elements (IRE). When it is bound, the mRNA is protected from degradation; when not bound (i.e. in the presence of iron) the mRNA is open to attack by RNA endonucleases, and is subsequently completely degraded into its constituent nucleotides by exonucleases.

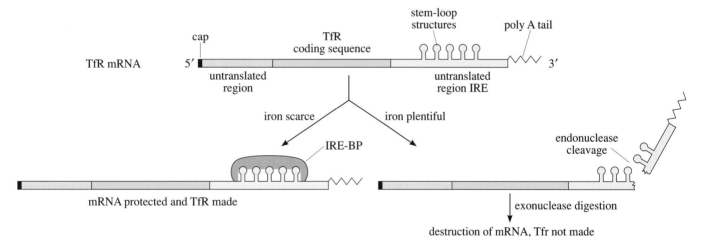

(b)

3.6 Translational control

The synthesis of a mature mRNA does not necessarily mean that it will be translated. Regulation of translation provides a further level of control. Translation is affected (and hence can be regulated) by several factors:

o The presence of appropriate initiation, elongation and termination factors (Section 3.1.2).

o The ability of ribosomes to attach to the mRNA.

o The longevity of the mRNA template (Section 3.5.3).

o The presence of enough of the appropriate tRNA molecules.

In mammalian reticulocyte cells (immature red blood cells) synthesis of globin proteins from mRNA takes place only in the presence of haem. Translational control is brought about by the haem-controlled repressor (HCR). When haem is absent, protein synthesis stops. The HCR, which is a protein kinase, becomes activated when the haem concentration falls and phosphorylates one of the translation initiation factors, eIF-2 (Figure 3.4). This factor, eIF-2, is inactivated by this phosphorylation and, as the initiation factor is an essential component of the translational apparatus, protein synthesis is inhibited.

▷ Why should the production of globin be linked to the presence of haem in this way?

▶ Haem is a cofactor which is required for the production of haemoglobin. Without this, haemoglobin cannot be formed and the globin proteins would be superfluous.

Returning to the example of the regulation of the genes involved in iron metabolism, *ferritin* synthesis is also regulated post-transcriptionally by iron. Ferritin is an iron storage molecule within the cell. In contrast to the TfR, ferritin synthesis is increased when iron is abundant. In the 5′ untranslated region of the ferritin mRNA is a stem-loop structure that resembles the stem loops of the TfR mRNA. The same protein that binds to the TfR mRNA iron-responsive element also binds to the stem-loop structure of ferritin mRNA. It is likely that this protein blocks ribosomal access to the ferritin coding sequence. Thus, the IRE-BP responds to decreased iron levels by:

(a) binding to the TfR mRNA 3′ end, protecting it from degradation and allowing transferrin receptor synthesis;

(b) binding to the ferritin mRNA 5′ end, preventing translation by blocking ribosome access and hence stopping ferritin synthesis.

Defects in the genes involved in iron metabolism are associated with haemoglobinopathies and anaemias. Haemochromatosis is an iron overload disease with many manifestations that arise from an abnormality of iron uptake and/or storage. The condition is inherited as an autosomal recessive. The gene responsible is known to be localized on the short arm of human chromosome 6, but has not yet been identified.

3.7 Post-translational regulation

Protein function and longevity may be affected by proteolytic enzymes which cleave the polypeptide chain at various places, as well as by kinases, methylases and other enzymes which modify particular amino acids. Some proteins require other modifications such as glycosylation before they are active. There are many examples of proteins that are produced as an inactive precursor that only becomes functional upon cleavage. For example, chymotrypsinogen is cleaved to produce chymotrypsin (which is coordinately expressed along with trypsin and pancreatic amylase in the exocrine (non-hormone-producing) cells of the pancreas).

Another example of such protein processing is in the production of the hormone insulin. The product of translation is a large precursor protein called pre-pro-insulin. First the amino terminal segment (which is a signal peptide specifying the destination of the molecule) is cleaved off and disulphide bridges are formed, resulting in pro-insulin. A central segment is then spliced out of the pro-insulin molecule to produce the active insulin hormone – two chains connected via disulphide bridges. Figure 3.26 illustrates all of the stages involved in producing a functional insulin molecule.

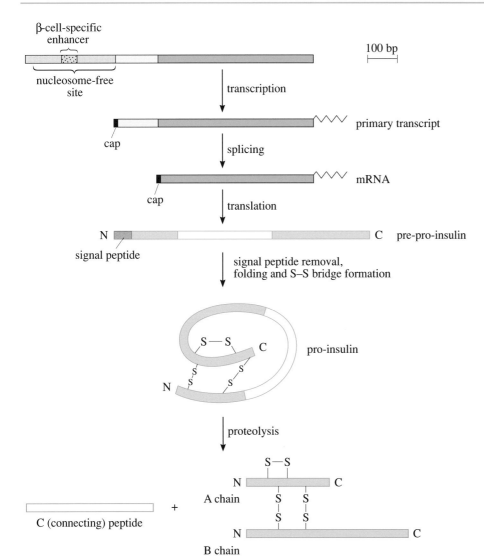

Figure 3.26 The stages involved in producing a functional insulin hormone molecule. This figure is a good summary of the principles described in this chapter.

3.8 Insulin: a case study of gene expression

This section is a brief overview of the role of insulin in the body, what happens when things go wrong (the causes and characteristics of diabetes) and how recombinant DNA technology has allowed the production of the synthetic hormone for pharmaceutical use.

As you saw in Book 3, hormones are chemical signals which travel throughout the bloodstream to influence target cells that are widely distributed in the body. Some, such as the steroid hormones are able to enter the cell and interact with their receptors and the DNA directly to regulate transcription of the target genes. For others, including insulin, the ability of a cell to react depends on the presence of receptors on the cell surface that bind the signalling molecule. In both cases, the ultimate effect is to alter the rates of synthesis of existing proteins or to initiate the synthesis of new ones.

Insulin synthesis occurs only in the β cells of the pancreatic islets; expression of the insulin gene is governed by *cis* acting transcriptional regulatory regions including a tissue-specific enhancer in the 5′ flanking region of the insulin gene. The major role of insulin is to control the utilization of carbohydrate. This is predominantly achieved by regulating the uptake of glucose into cells and its subsequent incorporation into the storage carbohydrate glycogen. In addition, insulin stimulates protein synthesis and, in fat cells, lipid synthesis. In this way, insulin signals the fed state. Insulin also acts to inhibit the release of glucose (gluconeogenesis) from stores in the liver. (It might help to look back at Figure 4.5b in Book 3.)

An increase in blood glucose levels stimulates endocrine cells in the pancreas to secrete insulin into the blood. Very quickly (within minutes) the increased insulin concentration stimulates liver and muscle cells to take up more glucose, thus reducing levels in the blood. The rate of insulin production is then reduced and hence the rate of glucose uptake by liver and muscle cells. In this way a constant blood sugar level is maintained. Insulin acts on the cell to increase glucose uptake by promoting translocation of glucose transporter molecules to the cell surface (Book 3, Chapter 3, Figure 3.11). Uptake is the rate-limiting step for glucose utilization.

The effect of insulin on the cells of the muscles and liver is dependent upon the presence of receptors on the outer surface of these cells. The insulin–receptor complex is endocytosed (taken into the cell) in vesicles and is broken down. However, the effects of insulin can be mimicked by specific antibodies that bind to the receptors on the surface of the cell. Therefore insulin itself is not the molecule that is the intracellular signal; *second messengers* (Book 3, Chapter 3) within the cell carry out this function. One effect of insulin is the phosphorylation of a single protein in the 40S ribosomal subunit (which effects control at the level of translation).

When there is a lot of endocytosis, in response to a large amount of hormone, the number of receptors on the cell surface decreases. This is because receptors are not produced rapidly enough to replace the dwindling stock. This is termed **receptor-down regulation**. In individuals who overeat, the blood sugar levels are persistently high. In response to this, insulin levels are elevated. The receptors are used up very quickly and the cells become relatively insensitive to insulin. The result is that the individual becomes obese. Most such individuals produce more and more insulin to stimulate the liver and muscle cells. Some individuals do not produce sufficient insulin and the resultant high levels of blood glucose may cause them to become diabetic.

3.8.1 Diabetes mellitus

You may like to refresh your memory by reading Book 3, Chapter 4, Section 2.5 before starting this section.

Diabetes mellitus, commonly referred to as 'diabetes', was recognized long ago in ancient Egypt. However, the biochemical features of the disease were only investigated in the last century. Diabetes is a chronic state of glucose intolerance which affects 2–4% of the population of the UK. The condition is characterized by persistent *hyperglycaemia* (elevated blood glucose) because the controlling effects of insulin are absent. Excess glucose is excreted in the urine when there is too much to be reabsorbed by the renal tubules in the kidney. This gives the urine its diagnostic sweet taste. The loss of glucose depletes the carbohydrate stores of the body, which leads to the breakdown of fat and protein. An untreated diabetic individual loses a lot of water in the urine as an accompaniment to the glucose and is constantly hungry and thirsty.

There are two forms of this condition: insulin-dependent diabetes (IDDM) usually has a juvenile onset; non-insulin-dependent diabetes (NIDDM) has a later age of onset. From studies with monozygotic twins (those arising from the same fertilized ovum and therefore having identical genetic material) IDDM seems to have both a genetic and an environmental component, whereas NIDDM seems to be solely the result of the genotype of the individual.

▷ How can twin studies provide this sort of information?

▶ If a phenotype is completely due to the action of genes, then two individuals with the same DNA (identical twins) should be the same. Thus if one twin suffers from a genetic disease, the other will also. However, there are often environmental factors involved and in these cases sometimes the twins will share the condition in question and sometimes they will not.

The chances of an individual suffering from IDDM are increased if they have certain genes in the major histocompatibility complex (Book 3, Chapter 3, and Chapter 7 in this book). Several disorders show **association** with certain alleles at the MHC locus. An **auto-immune** disease is one in which cells of the immune system are triggered by an as yet unknown stimulus to attack the cells of the body. In the case of IDDM, the β cells of the pancreatic islets are attacked. The resulting lack of insulin forces the tissues to use fuels other than glucose, with life-threatening metabolic consequences including renal failure and blindness.

The only treatment for IDDM is daily administration of insulin, without this the patient will die. However, an overdose of insulin can also have terrible consequences. Too much insulin results in *hypoglycaemia*, the symptoms of which include sweating, tachycardia (abnormally rapid heartbeat), hunger, mental confusion and coma.

▷ How might these conditions be alleviated?

▶ By administering glucose to reverse the effects of having too little glucose in the bloodstream.

Despite the large genetic component to NIDDM, there is clearly a link with obesity and, in some instances, eliminating obesity can remove all trace of the diabetes. NIDDM is the result of a number of predisposing genes rather than being a single-gene disorder (multifactorial inheritance). The plasma concentration of insulin in this condition is often normal but there appears to be an insensitivity to insulin in peripheral tissues, possibly as a result of a decrease in the number or affinity of receptors for insulin in the target cells.

Several different causes have been identified that lead to a lack of response to insulin:

(a) Mutations that alter the residues at the junction of the A or B chain and connecting peptide of pro-insulin can interfere with the processing into mature insulin. Some diabetic patients have a high plasma level of pro-insulin, which is not an active hormone.

(b) An amino acid substitution in a critical region of the molecule may affect its function. The replacement of a leucine residue by a phenylalanine near the C-terminus of the B chain reduces the activity of the hormone by 90%.

(c) Some patients secrete normal insulin but have defective receptors and so insulin cannot bind.

(d) Another group of patients have normal insulin molecules and normal receptors but the cells still do not respond.

▷ What could be the explanation for the insulin resistance in this last case?

▶ The likely defect is in the intracellular transmission of the hormonal signal – that is, a defect further along in the signalling pathway.

The symptoms of NIDDM are usually less severe than those of IDDM and are either treated by diet, or by oral hypoglycaemic agents which increase transcription of genes encoding glucose transporter proteins.

3.8.2 Recombinant insulin

Insulin is a medically important molecule. As you may recall from Book 1, it was the first protein to be sequenced, back in the 1950s. This provided a key to understanding the structure and function of proteins since before this time it was not certain that proteins had a definite, genetically determined sequence.

Insulin was also the first genetically engineered or *recombinant* drug licensed for human use. Prior to this, insulin was obtained from the pancreases of pigs and cows. Pig and cow insulin molecules are functional in humans but do have slightly different amino acid sequences. Some patients produced antibodies to this 'foreign' insulin which occasionally resulted in severe immune reactions. The safe production of sufficient quantities of insulin for the treatment of diabetes was a triumph of genetic engineering. The insulin gene is cloned into a plasmid that will replicate in *E. coli* and is placed under the control of a promoter in the plasmid β-galactosidase gene (*lac Z*) – that is, the the insulin gene is inserted within *lac Z* (Figure 3.27). The *E. coli* transcription and translation machinery then produce large quantities of the protein, which can be easily harvested. However, the scheme is not as straightforward as it appears. First, as we have seen, eukaryotic genes contain introns, whereas prokaryotic genes do not. Thus, since they lack the mechanisms required for splicing, bacteria cannot produce the correctly processed mRNA from a eukaryotic gene. Second, the insulin polypeptide requires processing as described above to produce the functional hormone. (Finally, many eukaryotic proteins are toxic to *E. coli* in large quantities or are degraded by bacterial proteinases, and so there are limits to the yields that can be achieved.)

The solution to the first problem is to use a cloned cDNA (Section 2.4.1) rather than a cloned gene, since in cDNA the introns are not present (remember that cDNA is a complementary copy of an mRNA molecule).

To get around the second problem, it is possible to use a cDNA corresponding to the final processed peptide. In the case of insulin, the two portions of the final polypeptide are produced separately in different bacterial populations, they are purified independently and then mixed together under conditions that allow the right disulphide bridges to form between the A and B chains (Figure 3.27).

Trials are still being carried out to study the effectiveness of these recombinant human insulin preparations. They are not problem-free, but they do appear to be chemically effective with low immunogenicity (that is, they do not tend to provoke an immune response).

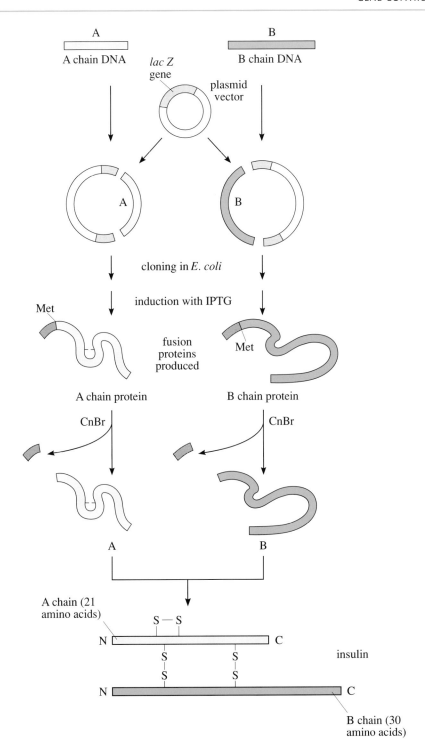

Figure 3.27 Diagram showing the production of insulin in *E. coli*. Synthetic cDNA molecules encoding the two insulin chains were synthesized and inserted into plasmid vectors under the control of the β-galactosidase (*lac Z*) promoter. The resulting proteins (fusion proteins) include a small region of the β-galactosidase peptide at the N-terminus and this is removed by cyanogen bromide treatment. The two peptides are then mixed together under conditions that allow the right disulphide bridges to form.

Objectives for Chapter 3

After completing this chapter you should be able to:

3.1   Define and use, or recognize definitions and applications of, each of the terms printed in **bold** in the text.

3.2   Recall the processes involved in transcription and translation.

3.3   Give reasons why control of gene expression is necessary.

3.4   Describe the features of gene control in prokaryotic systems – that is, the operon.

3.5   Illustrate the differences between eukaryotic and prokaryotic gene expression.

3.6   Describe the various stages at which eukaryotic gene expression can be regulated.

3.7   Appreciate the complexity of metabolic processes and the effect that disruption of gene control can have upon them as demonstrated by the various causes of diabetes.

Questions for Chapter 3

Question 3.1   (Objective 3.2)

Describe the roles of the three main types of RNA (mRNA, tRNA and rRNA) during translation.

Question 3.2   (Objective 3.2)

In what ways do pre-mRNA and mRNA differ?

Question 3.3   (Objective 3.4)

What are the main features of the control of prokaryotic gene expression as exemplified by the lactose operon?

Question 3.4   (Objectives 3.2 and 3.5)

How do the structures and features of transcription differ between prokaryotic and eukaryotic genes?

Question 3.5   (Objective 3.6)

What are the main features of transcriptional control of eukaryotic genes?

Question 3.6   (Objectives 3.1 and 3.6)

What are transcription factors? What types of protein structure might a transcription factor contain?

Question 3.7   (Objectives 3.3 and 3.6)

Summarize by means of an appropriate table, the iron-dependent regulation of the transferrin receptor and ferritin genes.

Question 3.8   (Objectives 3.1 and 3.6)

Why are tortoise-shell cats always female?

Question 3.9   (Objective 3.7)

For what reasons might an individual become diabetic? How can diabetes be treated?

4.1 Change and decay

Without variation, there could be no genetics and no evolution. This chapter is an account of **mutation**, the dynamo that generates inherited diversity. Nothing in genetics has changed more under the influence of technology than has the image of mutation. For many years, mutations were seen as rare and generally harmful events, errors in the workings of a finely adjusted machine. Now we know that sections of the genome are in constant flux. For some of our DNA at least, mutation is a common event and the concept of the '**fluid genome**', with change as one of its intrinsic properties, is central to the understanding of its structure. However, other regions of the genetic material appear to be much more resistant to the action of mutation and the balance between stability and fluidity in DNA structure is an area of controversy. We shall review the variety of mutations and how they can be detected, taking into account that some are lethal and so disappear before they can be counted, and that others are recessive and show their effects only when homozygous and long after they arise. Molecular biology has helped; changes in DNA sequence can now be identified as they appear. Chromosomal and molecular mutations have, it seems, a great deal in common: they can result in local alterations in structure or larger-scale damage or rearrangement of the genetic material. Both in the visible phenotype and at the molecular level, many mutations are due to interference by third parties (including, for example, mobile DNA which hops around the genome, causing damage as it goes). Finally, we shall look at **mutagenesis** – the generation of mutation by external influences such as radiation and chemicals – and how the body copes with constant genetic damage by mobilizing a complicated series of repair mechanisms.

The origin of life can be dated back to at least 3 000 million years ago. A key step in the development of living processes was the capacity of primitive cell-like structures to divide and replicate. This process itself came to depend on the ability of macromolecules (such as the nucleic acids DNA and RNA) to copy themselves accurately. If that were all they could do, and if they did their job perfectly, then life today would be a 'soup', just as it was then. Evolution needs mistakes, the raw material with which it works. Genetics too, is nothing but the study of inherited diversity. The world is full of variation, manifest both in the existence of millions of species and in the mass of heterogeneity which each one contains. All of it has arisen through inaccuracies in the copying of the hereditary material. This chapter deals with mutation; errors in the replication of genes.

▷ What kinds of errors might there be in DNA replication?

▶ There are two general classes of error; those involving changes in the DNA bases themselves, and those arising from damage to the sugar-phosphate backbone (which can lead to the loss of sections of DNA).

Inheritance is, in general, stable. If it were not, then Mendel's laws would not exist. However, its constancy is not perfect. Inaccuracies occur both in germ cells – sperm and egg – and in the somatic cells of the body itself (the soma). The boundaries between mutations in the germ line and in the soma are becoming blurred, and it is clear that many diseases – such as cancer – arise through **somatic mutations**. Ageing, too, may be at least in part a reflection of the decay of the genetic message in body cells. Some biological systems (such as the cell surface antigens that parasites use to combat the mammalian immune system) undergo frequent somatic mutations as part of their normal function. These may themselves be influenced by inherited instability in the genes involved. Errors in somatic cell replication will be explored further in Chapter 7: this chapter is concerned with germ-line mutation.

▷ Which kinds of mutation are more important in evolution?

▶ Germ-line mutations, as they are passed to subsequent generations. Somatic mutations, in general, affect only their carriers.

Mutations were once seen as rare and generally regrettable events. Most, it seemed, marred the perfection of a stable genotype. Now we know that, for much of the genetic material, this is far from true. The study of mutation has given a new insight into the structure of the genome, of why it looks the way it does and, perhaps, of why a large proportion of the DNA of eukaryotes has so little apparent function.

The study of mutation is important in research on evolution, on cancer, on ageing and on inherited disease. Because there are so many genes (and so much non-coding DNA) mutation is common in the genome taken as a whole. However, until recently it was one of the least understood aspects of eukaryotic genetics, largely because mutational changes are, for most individual genes, rare.

Darwin was well aware of their importance. In *The Origin of Species* he talks of 'sports' or sudden changes in a family from one generation to the next. He was particularly impressed by the Ancon sheep, animals with unusually short legs which arose – by mutation, as we would now say – in a normal flock. They quickly became popular because they could not leap over stone walls.

The first real hints about the changeability of genes came soon after Mendel's laws were rediscovered. The Dutch botanist Hugo de Vries suggested that differences among living species must have arisen by sudden large changes and not by the process of gradual evolution. He coined the term 'mutation' to describe them. He found many examples of sudden transformations of colour and structure in succeeding generations of the evening primrose *Oenothera lamarckiana*. Although we now know that this plant has an unusual system of chromosome instability which leads to an atypically high incidence of genetic change, his term has stuck.

▷ What unites both Darwin's and de Vries' archetypes of mutation and differentiates them from many others?

▶ Both are *dominant*: they manifest their effects as soon as they appear. In contrast, many human mutations of clinical importance are recessive and show their effects only many generations after they have taken place.

4.2 Studying mutations

Thousands of mutations have been discovered. Some are dominant, some recessive. They can involve changes in complete chromosomes, in long segments of chromosomal material, in short lengths of DNA, or in individual bases. Some cause obvious alterations – an extra pair of wings in *Drosophila* or shortened legs and arms in humans. Many mutations are **lethals** – they kill their carriers. Others are more subtle and show their effects only when placed in the appropriate environment.

▷ Can you think of an example of a mutation that damages its carriers only in certain circumstances?

▶ Think of the sickle-cell mutation. Homozygotes for the mutation may be quite safe until they get a fever, or find themselves in conditions where oxygen demand is high (see Chapter 1).

In bacteria many mutations remove the ability to make essential constituents of cells (such as amino acids) from simple precursors. Their presence becomes obvious only when the bacteria are placed on a medium which lacks the amino acid.

Such **conditional mutations** are found in higher organisms, too. Sometimes (as in the inborn disease phenylketonuria) their effects depend on the food available. Children with this disease have a mutation in a crucial enzyme that breaks down phenylalanine, an amino acid found in most foods. If they are diagnosed early enough (and, in developed countries at least, all children are now tested at birth), then the treatment is simple and effective. A special diet lacking phenylalanine minimizes the effects of the mutation.

▷ What chemicals might be particularly important in uncovering previously unidentified conditional mutations in humans?

▶ The effects of certain drugs are more severe if the individual being treated carries a mutation in one of the enzymes responsible for its metabolism. For example, the muscle relaxant *suxamethonium* – widely used in preparation for surgery – can cause those carrying a mutation in the acetylcholinesterase enzyme to become dangerously ill because administration of the drug results in paralysis of the breathing muscles. Screening populations for inherited differences in sensitivity is now an important aspect of drug development.

Other mutations depend on temperature. The eyes of the *Drosophila* eye mutant *white-blood* are brownish (rather than the wild-type's bright red) when the flies are raised at 20 °C, but almost white when the flies are raised at 28 °C, as the damage due to the mutated gene is more severe at high temperatures. Surveys of new mutations in *Drosophila* and other creatures suggest that as many as half may be temperature-sensitive.

▷ What sorts of mutations are likely to be temperature-sensitive?

▶ Any that weaken the physical structure of a protein (such as those involving an amino acid like cysteine, important in the maintenance of the higher-order arrangement of the molecule). Injuries to such sites are likely to be more obvious under conditions of high thermal stress.

Many mutations are, at least at first observation, **silent**, i.e. they have no apparent effect on the phenotype of those who carry them. Others are **neutral**: their effects may be manifest but they are so minor as to have no impact on the carriers' ability to survive and reproduce. This is likely to be true for genetic changes in the third position of the triplet code – UCA and UCG, for example, both produce serine (see Chapter 2, Table 2.3) and so cannot be differentiated without actually sequencing the DNA. Because so much of the DNA does not code for an amino acid, most mutation at the molecular level may be effectively silent in its effects on the carrier's phenotype and neutral in evolutionary terms.

▷ Within a gene, what sections are most likely to accumulate mutations?

▶ Introns are more likely to accumulate such mutations because changes in the DNA sequence have few effects on the workings of genes. Exons may be damaged by a new mutation, which will then rapidly disappear from the population.

4.2.1 Measuring the rate of mutation

There has been plenty of mutation during the history of life – after all, the difference between the giraffe and the geranium can, ultimately, be traced to it. However, simply counting the genetic differences between species, between members of the same species, or between parents and offspring does not measure its rate. For an altered gene to be detected it must persist: simply looking for hereditary changes thus conflates the rate of mutation with the effects of a new mutation upon the chances of its carriers surviving and reproducing.

There are plenty of cases in which a good guess can be made at just where and when a mutation first arose. The famous melanic mutation in the peppered moth *Biston betularia* was first recorded in Manchester in 1848 and rapidly spread in the newly polluted environment. It probably arose very shortly before it was first detected. The same is true of the mutations that conferred insecticide resistance on mosquitoes and house-flies when DDT was first used. Patterns of variation in the DNA close to the mutated locus suggest that a single genetic change in one individual gave rise to a mutation so advantageous that it quickly spread in subsequent generations until billions of flies, world-wide, became resistant. All resistant insects, wherever they come from, share the same short segment of DNA around the resistance gene, suggesting that all possess the same copy of the changed gene. Highly advantageous mutations such as these can be counted as they arise.

▷ Do all highly advantageous mutations arise just once?

▶ Probably not. Think of the separate foci of sickle-cell disease associated with distinct DNA haplotypes in different places (Chapter 1). Each of these probably represents a separate occurrence of the sickle-cell mutation, which quickly spread.

However, because not all mutations are advantageous, life is not always so simple. Most mutations arise and disappear with no evidence of their passing. Consider, for example, the structure and function of fibrin, active in the chain of biochemical reactions involved in blood clotting which, as we shall see later in this chapter, is a classic system for studying human mutation. A clot is largely made up of a tangled mass of interlocking fibrin molecules. Fibrinopeptides are lengths of amino acids whose job is to stop blood clots being formed at inappropriate moments. The fibrinopeptides of, say, humans and cows both do the same undemanding job. After a cut, the fibrinopeptide drops off the main part of the molecule and clotting proceeds. There are many amino acid differences between the human and the bovine fibrinopeptide molecule. Mutations have accumulated at the rate of about one per million years since the evolving lines diverged, which might, perhaps, be taken as an indication of the rate of mutation at this gene locus. Fibrin itself, though, is almost identical in cow and man. This is not because fibrin has a far lower mutation rate than does fibrinopeptide. As fibrin must bind with its neighbours to do its job it cannot accept any mutational changes. Only the harmless mutations in fibrinopeptide remain to be counted (see Figure 4.1). Looking at divergence – between species, individuals or generations – combines the rate of mutation with its effects: if the changed gene is less efficient than the one which went before, the new mutation will disappear.

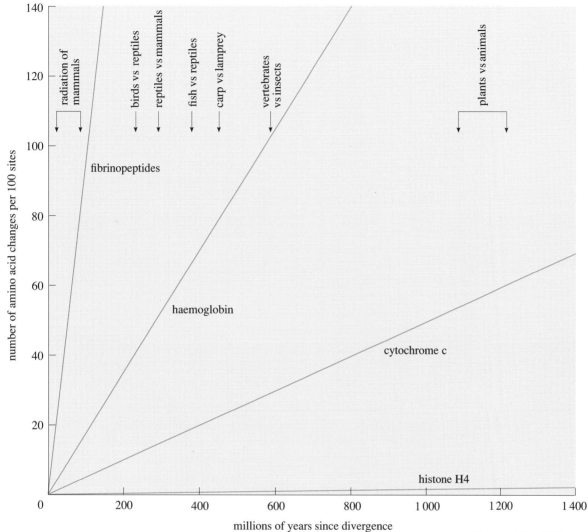

Figure 4.1 Differential accumulations of mutations in different proteins. Those with a crucial functional role (such as histones) show fewer mutational changes than those without.

The same problem arises in studying the rate of mutation over shorter periods of time. Mutations with a dramatic effect are easy to detect, but those with a less immediate impact are hard to find. For this reason, mutations seemed to be far rarer events 30 years ago than we now know them to be. There were two problems: thousands of mutations with no discernible effect on the phenotype were missed because there was no way of detecting them, and thousands more killed their carriers early in development before their presence could be identified.

There are other practical problems in counting mutations. The most important is that many new mutations do not make their presence manifest for many generations.

▷ What type of new mutations are unlikely to be detected as they arise?

▶ Those giving rise to alleles recessive to wild-type, whose phenotypic effects are only obvious when they appear as homozygotes, which may be many generations later.

All this means that estimating the mutation rate is not as easy as it might at first appear. The history of the study of mutation is rather like that of genetics as a whole: advances in technology have led to rapid shifts in views of how things work. The wheel has come full circle. De Vries simply counted changes in flower colour as they arose. There was then a period of elegant experimentation in which the rate of mutation was inferred from breeding experiments. Genetics is now entering a new age of counting mutations as they arise, an age made possible by molecular technology.

For a mutation to be identified it must have an observable effect. In diploids, therefore, new recessives must be made homozygous before they can be counted. Much of the early history of mutation research is an attempt to get round this problem. Since then, the difficulty has been circumvented by looking directly at the DNA of mutated genes. The problem of recessivity then disappears.

H. J. Muller first solved it in an ingenious series of breeding experiments on *Drosophila* which began in the 1930s and won him a Nobel Prize. He studied a class of recessive changes which can easily be identified because they are lethal when homozygous – they kill their carriers. His main interest was in the effects of radiation on the genetic material, but his approach can be used just as easily to measure the background rate of mutation.

It depends on two useful virtues of *Drosophila*. First, there are many accurately mapped and easily identifiable genetic changes which can be used to track the passage of a particular chromosome through the generations. Second, *Drosophila* chromosomes have **inversions** or rearrangements of the genetic material which prevent crossing over between an inverted and an uninverted segment of chromosome. This means that long sections can be held together for many generations without the linear arrangement of genes being disrupted by recombination.

Figure 4.2 shows Muller's scheme for identifying new lethal mutations. It detects such mutations on the X chromosomes. Exactly the same approach can be used, with a slightly more complicated marker stock, to study the rate of mutations on the autosomes.

The stock carries alleles that produce visible mutations at two loci, held together on an inversion. One, *Bar eye*, alters the shape of the eye – homozygotes have the eye reduced to a slit, heterozygotes to a kidney shape. The other, *apricot*, is a recessive eye colour mutation. The system holds the two loci (along with the many other gene loci on the same chromosome) together without recombination.

Imagine that a new recessive mutation has taken place on the X chromosome of the male being tested. It does not, of course, show its effects in the fly within which it actually occurred. Rather, it modifies the cells that give rise to its sperm. As all those inheriting that X-chromosome are female heterozygotes, it will also be concealed in the next generation. However, in the F_2 generation, the mutated X chromosome appears in hemizygous condition (that is, without a masking X chromosome) in males and, for the first time, shows its effects.

If the new recessive is a lethal (as many are) then those males will die long before they become adult. Because each of the four phenotypic classes expected in the F_2 generation can be identified from the appearance of their eyes and wings, stocks with new lethals stand out from the others because they lack red-eyed males with normal wings. The rate of mutation can be worked out by repeating the experiment thousands of times and counting the proportion of vials which lack wild-type males. The availability of stocks analogous to Muller's

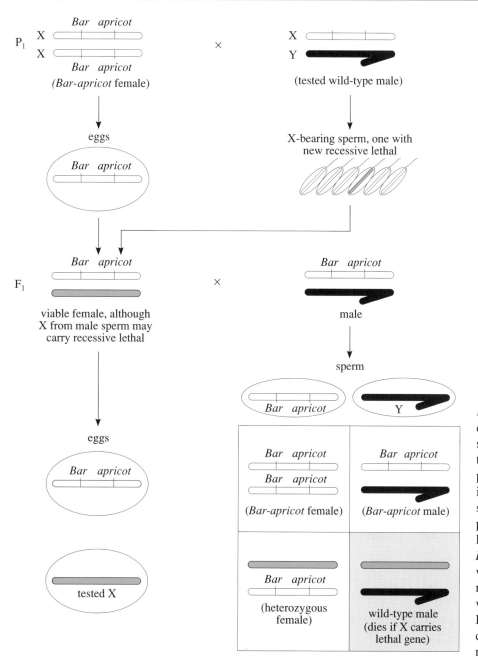

Figure 4.2 The mating system used to detect recessive lethals on the X chromosome. The wild-type parental male being tested may, as a result of mutation, produce many kinds of X chromosomes in his sperm, some of which bear recessive lethal genes. An F_1 female is protected against sex-linked recessive lethals by the presence of the lethal-free *Bar-apricot* (Muller-5) chromosome and will produce *Bar-apricot* and wild-type male offspring in equal proportions if the wild-type X chromosome is fully viable. However, if the wild-type chromosome carries a recessive lethal, no F_2 wild-type males will appear.

F_2 males expected: 1 *Bar-apricot*, 1 wild-type

F_2 males if X carries recessive lethal: all *Bar-apricot*

Bar-apricot stock for each chromosome (and even for short lengths of individual chromosomes) means that *Drosophila* is still widely used in the study of mutagenesis. As we shall see later in the chapter, this approach was of crucial importance in studying the effects of radiation on the genetic material.

▷ Which chromosome should, in principle, show most easily the effects of new recessive mutations?

▶ The X chromosome because it frequently resides – in single copy – in males, exposing the effects of all gene substitutions, dominant or recessive.

It was once hoped that a similar procedure could be used to study mutation rates in humans. Perhaps there would be a reduction in the number of male children after a population had been exposed to an agent that might produce mutations. Any new lethal mutations on the X chromosome would manifest their effects immediately on the sons of those exposed. Daughters would be safe, as recessive lethals would be hidden by a normal allele on the other X chromosome. This neat idea did not work.

Muller's experiment summed the effects of mutations at many loci on the chromosome simultaneously. After all, a fly needs to be a homozygous lethal at only one locus to be dead. This is not particularly important when looking at the net rate of mutation over a long section of chromosome or comparing – say – different radiation doses for their mutagenic effect, but the technique is rather a clumsy one to use when trying to study mutation at individual loci.

One way of looking for mutations at single loci is with **tester stocks**: inbred lines homozygous for recessive alleles at a series of loci with detectable effects on, say, colour. When these are mated with wild-type individuals (those with two copies of the normal, dominant allele) all their offspring will, of course, be heterozygotes at each locus and will have the wild-type phenotype – unless, that is, the germ cells of the individuals being tested have undergone a mutation. If a mutation from wild-type to the allele present in the tester has taken place, then the phenotype of the offspring (now a homozygote recessive at that locus) will resemble that of the tester stock.

This approach has been much used in mice. The tester stock is homozygous at several loci for recessive alleles producing coat characters such as 'leaden', 'brown' and 'dilute'. If this stock is mated with wild-type mice, then their offspring are heterozygous at all the loci involved and themselves appear wild-type. However, if there have been any mutations in the wild-type mice to a recessive allele at one of the tester stock's loci then offspring with mutant coat colours will appear. Tens of thousands of mice must be examined, but the investment has paid dividends for understanding mutation in mammals.

4.3   A direct approach: the New Age of counting

Inferring the rate of mutation by breeding experiments is a slow and tedious process, at least in organisms with lengthy life cycles. A direct approach – looking at new mutations as they arise – has always enticed geneticists. Until recently, this approach has only been feasible in simple creatures, such as bacteria. Now, though, it is giving a new insight into the nature of mutation in humans.

At first consideration, it seems simple to count new mutations as they arise. In humans, any dominant mutation should surely be easy to identify since the offspring look different from either parent. Many human mutations must arise *de novo* as neither parent shows their effects. The majority of children with dominant achondroplastic dwarfism (which interferes with the growth of the arms and legs), for example, are born to normal parents. A mutation must have occurred in the germ line of one of the parents. The same is true for any dominant allele which prevents its carriers from reproducing (often by killing them at a young age). As the allele cannot be transmitted, the mutation rate must be the same as its frequency in the population.

However, this is not an accurate way of measuring the overall rate. Many new mutations are lethal early in pregnancy and are missed: one estimate is that nine fertilized eggs out of ten are lost because they carry new dominant mutations (many involving chromosomal changes). In addition, what might seem to be a simple condition under the control of a single gene – inherited deafness, for example, – may in fact reflect damage at several independent loci. This means that the rate of mutation at a single locus may be exaggerated.

▷ What other characters might show similar phenotypic effects arising from mutations in independent genes and hence confuse the process of counting genetic mutations by looking at phenotypic change?

▶ Any metabolic pathway involving several enzyme-catalysed steps, any one of which might be blocked by mutation. Many bacterial metabolic pathways are of this kind, as is, for example, the pathway leading to human melanin production (which explains why there is more than one genetically distinct form of albinism).

There are more mundane problems, too. Certain conditions may arise either from mutation or from an environmental problem that disrupts the process of development. This was true in the immediate aftermath of the thalidomide disaster of the 1960s. Many pregnant women had taken the drug. The first few children born with phocomelia (reduced arms or legs) were thought to carry new mutations until the number of cases became so high that it was clear that something else must be involved. Other apparent mutations may also arise from environmental causes, and, if this is missed, the rate of mutation will be overestimated. It is also sometimes the case that the biological father of a child is not the marital father who appears in a genetic counselling clinic. An apparently new allele in the child may come from his or her real parent rather than from a new mutation. All this means that early calculations of the human mutation rates are higher than those now generally accepted.

Direct counting of human mutations as they happen is, nevertheless, feasible in principle. All that is needed is an unambiguous method of comparing the phenotypes of parents and children that avoids the problem of recessivity. The first – and largest – survey was carried out on the populations of the cities of Hiroshima and Nagasaki. It compared the genotypes of children with that of their parents for loci coding for blood proteins. As we saw in Chapter 2, electrophoresis separates biological molecules on the basis of charge and size. This can be done just as well for proteins as for DNA.

In the Hiroshima and Nagasaki study, a modified technique, **two-dimensional electrophoresis,** was used. The gel is turned through 90 degrees after the initial operation, the proteins denatured, and the current applied again. The proteins are smeared out for a second time and, when the right stain is applied, thousands of small blobs appear. Each blob represents a separate gene product (whose actual function may be quite unknown).

Mother, father and child are all examined and a computer used to compare the images. If one of the blobs has shifted to a new position in the child's results compared to those of its parents, a new mutation must have occurred. This method did reveal some genetic alterations in the populations of the two cities but, as we shall see, has reopened questions about the dangers of radiation (see Section 4.7 below for discussion).

Figure 4.3 shows examples of two-dimensional protein profiles obtained using this technique (in this instance from bacteria rather than humans).

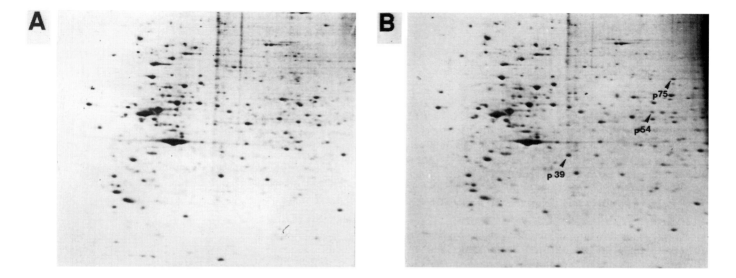

A

B

Figure 4.3 Two-dimensional protein profiles of (A) uninfected, and (B) infected cells. The arrows in (B) indicate the blobs that represent proteins not present in the uninfected cells. Although to the naked eye the profiles appear difficult to assess, modern computer systems and staining techniques allow accurate comparison of genotypes.

This approach, powerful though it is, does not identify what the mutated locus might do, and looks only at gene products rather than the DNA itself. Technology has at last made it possible to examine actual changes in the genetic material. The information that is beginning to emerge is changing our whole view of what mutation is and how it takes place.

Sequencing the DNA of patients to identify the nature of their mutation is still a slow business. There are various technical tricks that help. If DNA is made into a single-strand, its bases attempt to pair together and the molecule folds up on itself. Even a single base change can alter this shape – and hence the rate at which the folded molecule moves through an electrophoretic gel. The search for these **single strand conformational polymorphisms** (SSCPs) in particular genes is a quick and easy way of searching for mutant alleles.

Once the exact nature of a particular mutation has been uncovered within the DNA of a single patient (who, for recessive conditions, will of course carry two copies of the damaged gene) it becomes possible to search for its carriers – heterozygotes who show no phenotypic effects – or for embryos who may themselves be homozygotes.

This process is now central to genetic counselling as it allows both carriers and homozygous embryos to be detected. A matching copy (an **allele-specific oligonucleotide**) of the crucial section of the normal and the mutant allele is made. The normal probe hybridizes more faithfully to normal DNA, while the probe based on the DNA sequence of the mutated gene has a stronger affinity for mutant DNA. The extent of binding can be assessed with standard methods. By synthesizing a range of allele-specific oligonucleotides to match each known mutation in a particular locus many individuals can be screened quickly. The technology of DNA sequencing is itself advancing rapidly (see Chapter 2) and this, the most direct method of all, is also being used to search for new mutations as they happen.

4.4   Kinds of mutation

All mutations involve a change in the DNA sequence or in the order of the DNA bases. Such events can take place in many ways – some simple, some complex and unexpected.

The first indication of just how complex mutations can be came from studies of chromosomes. Although chromosomes are much more than magnified versions of DNA molecules, variations in their structure mirror what is happening, on a much smaller scale, in the genome itself. Chromosome mutations have the great advantage that many of them can be detected by looking down the microscope, particularly with new staining technology.

4.4.1 Changes in chromosomes

The most noticeable chromosomal mutations involve the gain or loss of whole chromosomes – or even of chromosome sets. Not surprisingly, they are often lethal. However, such events must often have happened in evolution, as related species may differ in chromosome number. In some groups, such as grasses and salmonid fishes, certain species have twice as many chromosomes as do their close relatives – they are **tetraploids**.

Most plants and animals are, of course, diploid: they have two copies of a chromosome set, one from each parent. The numbers of multiples can change, by mutation, to triploids (3*n*), tetraploids (4*n*), and so on. **Polyploid** plants and animals are common. Often, natural polyploids arise from the incorporation of the chromosome set of one species into that of another. This has happened extensively in crop plants. Wheat, for example, incorporates genomes from at least three wild species of grass.

Sometimes, because of the ensuing difficulties of meiosis, polyploidy can lead to sterility. This has been used by plant breeders in such commercially attractive breakthroughs as the development of the seedless melon. Animal polyploids, too, may be forced to abandon sex. There are in the deserts of California thousands of small lizards, *Cnemidophorus*. No male has ever been found. This species is a polyploid, and reproduces *parthenogenetically*, females making genetically identical copies of themselves without fertilization.

Multiplication of the number of chromosome sets in humans is invariably fatal. Very occasionally a child is born with a mixture of diploid and triploid cells (probably because of a somatic mutation, an error in a dividing body cell). It is a **mosaic**, possessing cells of different genotype. Such children die soon after birth. Indeed, the majority of pregnancies end in spontaneous abortion, usually very early on and before the woman herself realizes that she is pregnant. Many of these failures are due to abnormalities of ploidy.

Although changes in ploidy are very damaging, changes in the number of copies of a single chromosome may have less devastating effects. An individual who has gained or lost a chromosome is an aneuploid who no longer has the correct dose of genetic material. If there is an excess chromosome, he or she is referred to as a trisomic (see Chapter 2).

Aneuploids arise from non-disjunction, a failure in the separation of chromosomes during the formation of sperm or egg (see Chapter 1, Section 1.4.2). Normally, a complete haploid genome is passed to one or other pole of the dividing cells during meiosis. Non-disjunction means that one gamete gains a chromosome, and the other loses. Once again, plants are much more likely to survive such errors than are animals.

As you learnt in Chapter 2, aneuploidies often happen in humans. With the exception of the sex chromosomes, a missing chromosome is invariably lethal. Trisomics, too, are very damaging. Some do survive to birth and beyond.

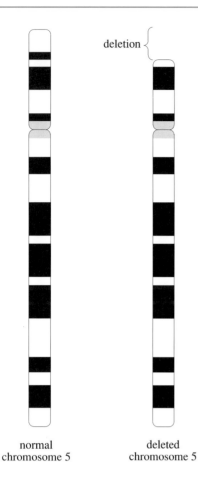

deletion

normal
chromosome 5

deleted
chromosome 5

Figure 4.4 The cause of the *Cri du chat* syndrome in humans is the loss of the tip of the short arm of one of the homologues of chromosome 5.

Trisomy 21 gives rise to Down's syndrome, one of the commonest of all inborn abnormalities (at about one birth in six hundred). Other aneuploidies include Turner's syndrome (XO, missing a Y chromosome and female) and Klinefelter's syndrome (XXY, a male).

Many mutations involve a gain or loss of chromosomal material. Sometimes a small section of chromosome is lost, giving a **deletion** (see Chapter 2). The disappearance of a large block of genetic material is often lethal. Occasionally, though, a cytological deletion does not kill its carrier, but changes its phenotype. The *Drosophila* wing mutant *notch*, for example, is the visible manifestation of a small heterozygous deletion which is, like most deletions, lethal when homozygous. In the heterozygote, the only effect is that the wing has a small piece scooped out of the rear edge.

Deletions have also been identified in human chromosomes. Again, most are lethal but some syndromes can be tracked directly to the loss of a small section of chromosome. *Cri du chat* syndrome (Figure 4.4; see also Chapter 2) is a rare, and invariably fatal, inborn disease. Affected children have a small head and a rounded face. They are mentally retarded. The disease gets its name from their characteristic mewing cry. Two bands on the tip of chromosome 5 have been deleted; a **terminal deletion.**

▷ Deletions removing an internal portion of the chromosome (**interstitial deletions**) are less common. Why?

▶ Because they require two breaks in the DNA rather than one.

Because deletions remove identifiable short sections of chromosome, they have been useful in gene mapping; an individual carrying a deleted chromosome may show the phenotypic effects of one or more normally recessive genes which have been unmasked by the loss of their wild-type alleles. This shows just where these loci must be. Human gene mapping has been greatly helped by the use of deletions (which commonly occur in cancer cells). If a probe for a particular gene fails to bind to a cell line carrying a specific deletion, then the locus involved must be somewhere in the deleted segment (see Chapter 2).

A **duplication** on the other hand involves the gain of chromosomal material. Duplications usually arise from errors in crossing over or as the consequence of other chromosomal rearrangements, such as inversions (see below). Sometimes there is the duplication of an entire chromosome arm.

Chromosome duplications contain within themselves a mechanism for increasing their number. If an individual homozygous for the duplicated chromosome appears, then – as can happen to zip-fasteners done up in a hurry – the pairing mechanism between chromosomes may slip. One chromosome with an extra copy of the duplication picked up from its partner chromosome can then emerge (Figure 4.5). Such "tandem duplication" has taken place in the *Drosophila Bar* mutant. This changes the shape and size of the eye. *Bar* homozygotes have a reduced, slit-like, eye.

Sometimes, homozygous *Bar* stocks produce a few offspring with tiny eyes and a few whose eyes are normal. The *ultra-Bar* flies have a third copy of this band, and those with normal flies just one. There has been a lopsided exchange during meiosis, some gametes gaining an extra copy of the *Bar* sequence and others losing one, returning to wild-type. Although, on the chromosomal level, mutation

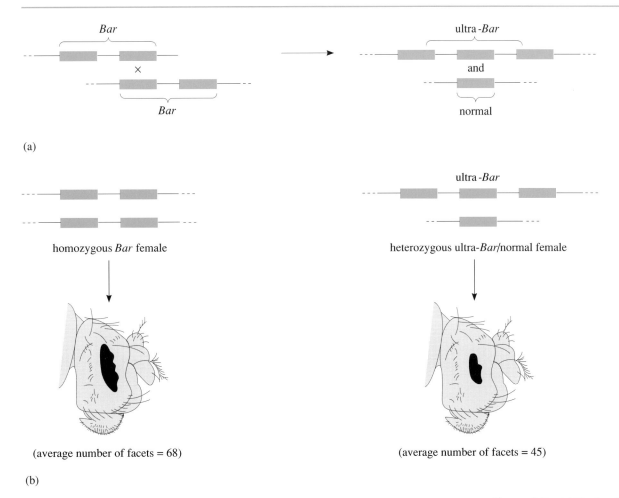

(a)

(b)

homozygous *Bar* female

(average number of facets = 68)

heterozygous ultra-*Bar*/normal female

(average number of facets = 45)

Figure 4.5 (a) Production of ultra-*Bar* (triplication) and normal chromosomes by asymmetric pairing followed by crossing over in a duplication homozygote. (b) Comparison of *Bar* and ultra-*Bar* phenotypes.

such as this (involving unequal crossing-over, see Chapter 2) is somewhat of a curiosity it may be of great importance in changing the sequence of DNA.

Changes in the number of copies of a tandemly-repeated sequence can also be caused by a process called **replication slippage**. In this case, the DNA polymerase may lose its place on the DNA and slip forward, thus missing out some of the copies. This mechanism is thought to be important in generating sequence diversity in the microsatellite DNA that you came across in Chapter 2.

Chromosomes can also mutate by rearrangement, without the gain or loss of material. Sometimes, sections of one chromosome shift to another one – a **translocation**. Often, the change is mutual; there is a **reciprocal translocation,** with exchange of material between two non-homologous chromosomes. Shuffling the genetic material in this way often happens during evolution. Some mouse populations isolated in mountain valleys have undergone many shifts of chromosome arms.

▷ Occasionally, children with Down's syndrome are born with 46 (not 47) chromosomes. How might this be explained in chromosomal terms?

▶ Such children do in fact have an extra copy of a section of chromosome 21, but this has been translocated to another chromosome in a previous generation. In these circumstances, Down's can become a heritable condition, rather than arising independently as a new mutation in each child born with the disease.

Inversions involve two simultaneous breaks in a chromosome, the rotation of the intervening segment, and its re-insertion into its original position. A chromosome in which the order of bands might be ABCDEFGH would be reconstituted as ABC*FED*GH. Although inversions (which do not change the total amount of genetic material) usually do not do much damage in themselves, they can interfere with the process of crossing-over during meiosis, and hence with recombination.

In inversion heterozygotes (and most newly arising inversions will, of course, be in heterozygous condition) each segment of one chromosome pairs, as usual, with the matching segment of its homologue. Because the order of genes along the part of the inverted chromosome is reversed, this pairing is difficult, for simple mechanical reasons. An **inversion loop** (see Figure 4.6) is formed as the chromosomes twist to make the pair. This reduces the opportunities for chiasma formation because of the mechanical stresses involved.

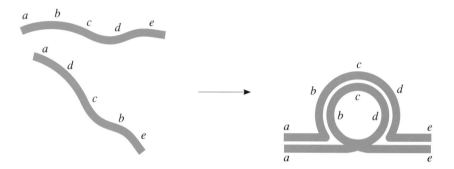

Figure 4.6 Inversion loop in paired homologues of an inversion heterozygote.

Another problem arises in individuals heterozygous for a **paracentric** inversion, one that does not span the centromere. The centromere is, of course, the site of attachment to the spindle during cell division. In a normal division each of the centromeres in a pair of sister chromatids passes to the opposite pole of the dividing cell. If a centromere is lost, meiosis is disrupted. The **acentric fragment** (the one without a centromere) has no guide to lead it to a daughter cell and is usually lost. The **dicentric fragment** has two centromeres, each pulling in opposite directions. It forms a bridge between the daughter cells and is broken as they pull apart. The broken sections each represent a chromosome with a deletion at one end.

As this is usually lethal, recombinant gametes emerging from a cross-over event within a paracentric inversion do not survive. The inversion is hence a **cross-over suppressor**; there is almost no recombination between loci within the inversion itself.

4.4.2   Mutational change at the DNA level

There is a wide range of structural mutations in chromosomes. They represent only part of the spectrum of mutations known to exist. For example, as we saw in Chapter 1, sickle-cell disease represents a single amino acid change in a polypeptide, an event far too small in scale to be seen down the microscope. Each chromosome mutation must, of course, represent a change in the DNA somewhere along the chromosome although where and what this might be was, until recently, quite unknown. Now that DNA changes can be examined directly, it is clear that the diversity of chromosomal changes is reflected in the complexity of mutation at the level of DNA sequence.

Gene mutation is much more than a microcosm of chromosome mutation. However, all the structural modifications seen in chromosomes have their counterparts on a smaller scale in the DNA. DNA is also susceptible to an assortment of other changes, some obvious and some surprising.

A mutation from one purine to another (or one pyrimidine to another pyrimidine) – C to T, or A to G – is known as a **transition**. The rarer shift between a purine and a pyrimidine (C to G, for example) is a **transversion**. The effects of such changes depend on where they take place. Should they occur in that large part of the DNA which does not code for a protein their phenotypic influence may be too small to measure. If they happen in a crucial part of a coding gene, a single transition may be lethal or may give rise to disease.

Even within the coding sequences of genes certain point mutations (that is, mutations within genes) have little apparent effect. These include 'silent' or same-sense mutations, in which the third position base of a codon is changed, but the normal amino acid coding is retained, and others in which one amino acid is replaced by another (mis-sense) which may or may not lead to a protein with an apparently similar function. Single base changes can, however, have major effects should they convert a coding triplet into a 'stop' codon abruptly cutting off the growing amino acid chain. This has happened in some haemoglobin chain-termination mutations and is termed a non-sense mutation.

▷ Using Table 2.3 in Chapter 2, list the codons which could, with a single base change, mutate to a STOP codon.

▶ The three stop codons are – in the RNA code – UAA, UAG and UGA. Codons such as UAU and UAC (both of which code for tyrosine) or UGU, UGC (cysteine) and UGG (tryptophan) can readily mutate to a stop codon.

Other mutations involve chemical changes to the bases. The link between a base and its deoxyribose partner may be broken, or there may be a spontaneous degeneration of one base into a related form. Cytosine breaks down to uracil at a rate so high that the function of DNA would quickly be disrupted if it were not for the presence of a series of specialized **repair enzymes** which reverse the process. Chemical damage to bases is, not surprisingly, greatly increased if reactive materials are present in the cell, as they often are. Hydroxyl ions, hydrogen peroxide and other high-energy substances are produced during respiration. All have the capacity to damage DNA.

Sometimes the chemical damage involves the formation of inappropriate matches within the DNA. For example, **thymine dimers** are pairings between two adjacent thymines which disrupt the replication of DNA.

Many mutations arise from deficiencies and duplications, in which sections of the genetic material – from a few to many thousands of base pairs – have been knocked out or multiplied. Their use in gene mapping has already been described. Such changes often happen in regions of the DNA that are rich in repeated sequences, suggesting that mistakes in the pairing mechanism during recombination (as in the case of the *Drosophila Bar* mutation) have increased or decreased the number of copies. Some parts of the bacterial genome are mutational hot spots (see Chapter 2), with a mutation rate 50 times higher than that of the surrounding DNA. Sequencing shows that these are, in fact, regions of repeated DNA and that many of the mutations arise from mispairing and the gain or loss of a short sequence. Even within human genes there are short sequences with a greatly elevated mutation rate. The enzyme phosphoglucomutase (PGM) for example, has a large number of distinct alleles distinguishable through gel electrophoresis of the protein. These seem to have arisen because of a region of genetic instability within the functional gene.

The segments of the genome that are analysed to create the DNA fingerprints used in forensic work have an astonishingly high rate of mutation (Chapter 2). The pattern of bands in the children are a mixture of those present in the parents. However, due to the high mutation rate, as many as 1% of children have a 'new' band that is not present in either of the parents. It is an altered form of one of the parental alleles. As you have seen in Chapter 2, the uniqueness of a DNA fingerprint lies in the difference in the number of copies and position of a short repeated sequence of DNA. Once again, inaccuracies in the process of DNA replication lead to a high rate of mutation.

4.5 Mutation by third parties – the role of mobile DNA

Mutation was once seen simply as damage, disfiguring an otherwise functional set of genes. Molecular biology has altered this view, and mutation is now thought of as being as much an inherent property of DNA as a response to external injury.

The change in the image of mutation – and of the genome as a whole – began not with sophisticated molecular biology techniques but with crossing experiments on maize. This plant is useful to geneticists as each cob is, effectively, a family of individual seeds all held together.

In the 1930s, an albino maize stock was discovered which had a very high rate of mutation: so high, indeed, that the plants developed as mosaics, individuals containing distinct clones of cells because of genetic accidents during development. Some of the clones contained mutations at a locus responsible for changing seed pigmentation. In an initially colourless stock the kernels were covered by dots of purple pigment, each dot representing a cell line that had mutated back to wild-type during development.

The American geneticist Barbara McClintock (who, much later, won the Nobel Prize for her work) set out to discover what was going on. Her crosses soon showed that the extraordinarily high mutation rate at the seed-pigment locus depended on another, unlinked locus. Only if a specific allele is present at this second locus does the colour locus show such a high mutation rate.

Several systems in which one gene increases the mutation rate of another are now known in maize. They have an additional peculiarity: mapping the mutator gene using conventional Mendelian genetics showed that it is at different chromosomal locations in different stocks and that it (and, in some cases, even its target locus) can change position in succeeding generations.

Ideas put forward to suggest that gene loci can wander around the genome, or that alleles at one locus can increase the mutation rate at another, were in their day sufficiently bizarre to be almost ignored. However, it was the first insight into what is now generally accepted; that genomes possess mobile sections of DNA that can copy themselves and insert their offspring elsewhere.

By making a cDNA – a matching copy of the actual DNA sequence – from the mRNA of the maize target gene and of its mobile element the structure of each has been analysed. The element is about 4 500 bp long, containing within itself the code for the enzyme – a **transposase** – which allows it to duplicate itself.

The original mutation in the tester stock (wild-type to albino, for example) is due to the insertion of an incomplete version of the element, a version that disrupts the working gene. Often, the transposase section has been deleted, so that the inserted section cannot move of its own volition. Only when a complete version is introduced into the stock by crossing does the partial element regain its mobility, sometimes hopping out of its temporary home and restoring that cell line's ability to make pigment.

Many such '**mutator genes**' – some of which increase the mutation rate at several loci simultaneously – are now known. Often, they show the eccentric genetic behaviour first found in maize: they can leap around the genome, sometimes appearing in one place and sometimes in another.

Drosophila melanogaster contains a whole range of transposable elements. They were discovered when flies from North America were crossed with those from other parts of the world and a sudden and, at first, inexplicable increase in the mutation rate at loci controlling morphological characters was observed. The effect was due to the mobilization of a **P element**, a piece of transposable DNA. Crosses between modern American stocks and those collected in the same place 50 years earlier and kept in laboratory culture since then also showed such **hybrid dysgenesis.** The older stocks lacked P elements, which must therefore have invaded *Drosophila melanogaster* quite recently. It now seems that they did so from a South American species of *Drosophila*. Often, transposable elements insert themselves into a working gene. DNA sequencing shows that many of the classic visible mutations of *Drosophila* are due to the insertion of one or other transposable element. Humans, too, possess sections of DNA that appear to have moved around the genome in the recent evolutionary past. The *Alu* elements discussed in Chapter 2 seem to have a history of this kind.

The study of mutational change at the molecular level has shown that what appeared to be a curiosity is, in fact, almost the rule: many mutations (including several in humans) result from the interspersion of autonomous genetic elements into working genes. Agents that increase the mutation rate often act by mobilizing such elements and they produce many mutations with an obvious phenotypic effect.

Unstable DNA is, it has been found, important in the genesis of some inherited diseases. It explains some previously baffling patterns of inheritance and also why some genes seem to be unduly prone to mutational change.

The commonest single cause of inherited mental illness is the fragile-X syndrome. Children born with the disease may be severely mentally retarded. They usually have a distortion of the X chromosome – two small pieces which seem to be breaking away from one end of its long arm (Figure 4.7). The X-linked nature of the defect means that boys are far more likely to be affected than are girls.

Studies of fragile-X families showed several puzzling features. Just as for any sex-linked locus, normal women may have affected sons. However, affected fathers can have affected daughters and, in those few families in which several generations are available for study, the severity of the disease can increase or decrease over two or three generations. What is going on?

The discovery that many mutations are due to mobile DNA gave a clue to the mystery. The locus involved was tracked down by the standard methods of mapping (see Chapter 2). The *FMR-1* gene, as it is called, codes for a protein that binds RNA in the cytosol and is particularly abundant in nerve cells – particularly

fragile X   normal X   fragile X   Y
(a)   (b)

Figure 4.7 A fragile-X chromosome, as seen in metaphase spread (Chapter 2), of (a) a heterozygous female and (b) a male, showing the small portion at the tip of the long arm that separates in suitable culture conditions (there is a connection, but it is not visible).

those in the parts of the brain thought to be involved in learning. It is a complicated gene, with 17 exons in its 38 kb length.

In the first exon is a region in which the triplet CCG is common. In normal people there are between six and 60 repeats of the sequence. Those with fragile-X syndrome have up to several thousand, and the more severe the symptoms, the more copies of the CCG repeat are present. This overabundance leads to an excessive methylation (see Chapter 2) of the *FMR-1* gene and transcription shuts down.

Some apparently normal people have **premutations**; between 60 and 200 copies of the triplet. These have no phenotypic effect, but when a woman with more than 100 copies has a son there is a high chance that the number will increase and that the child will be affected. Only when the gene is transmitted through females does this expansion take place. In the few instances in which a fragile-X male has a child the number of copies actually goes down.

This new form of mutation helps explain the strange genetical phenomenon of **anticipation**; described many years ago, but largely forgotten. Within a lineage, the symptoms of a disease get worse in each generation, an effect at least partly due to repeated mutation and the accumulation of short segments of DNA. It is involved in several other inborn errors. Huntington's Disease (HD) – a dominant nervous degenerative disease with variable age of onset – is due to the multiplication of copies of a CAG repeat within a working gene. Those with the disease have more copies of this triplet. The crucial number is around 40 repeats: below 34 there are no symptoms (and the repeat is found in normal individuals), but should the number rise by mutation to above 42 then the disease is liable to show its effects (Figure 4.8).

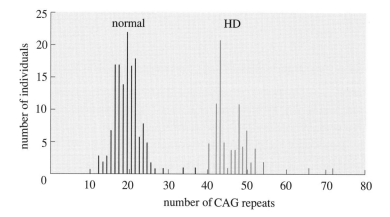

Figure 4.8 Distribution of allele size on normal and HD chromosomes. 145 independent normal chromosomes and 85 HD chromosomes were derived from 36 HD families.

Huntington's is a disease whose first insidious symptoms – often no more than depression and a generalized malaise – usually appear in middle age, and gradually get worse, often killing the patient. The disease and the families it affects are discussed in Video Sequence 4 along with the techniques used for genetic analysis.

The average age of onset is at about 38 years. However, this is very variable.

▷ How might the age of onset vary?

▶ The age at which the symptoms show themselves depends on how many copies there are (see Figure 4.9).

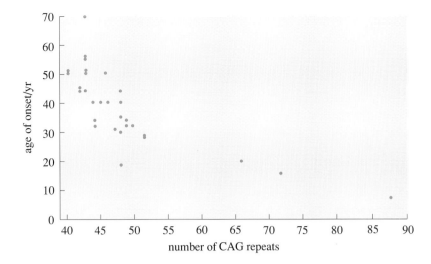

Figure 4.9 The relationship between number of CAG repeats and age of onset of Huntington's disease. Age of onset in 30 HD patients is shown as a function of the repeat number.

This time, the number of repeats is more likely to increase if the gene is passed through the father than through the mother. More than a third of all cases of the disease in which the gene was paternally transmitted show an increase in the number in child compared to father – an astonishing rate of mutation (for that is what it is). Sometimes, the number of repeats leaps by up to 40 in a single generation of transmission. Children with large numbers of repeats (80 and more are sometimes found) may show symptoms before their tenth birthday. An increase in copy number through this novel form of mutation in this highly unstable gene is hence clinically important.

Myotonic dystrophy, a sex-linked recessive muscle degenerative illness, shows the same fluidity, and for the same reasons. There is even instability in the number of repeats in somatic cells. Some patients are mosaics, with different numbers of copies of the repeat in different cells. Again, mutation appears to be a much more dynamic process than once imagined. Mutation by increased repeat number seems to be particularly prevalent in the genes controlling nerve and muscle; why is not known.

4.6   Haemophilia: a microcosm of molecular mutation

The intricacy of mutation is best illustrated by looking at a single gene and its errors. Appropriately enough, the system that best illustrates human mutation is the most famous inborn error of all – the disease that affected many of Queen Victoria's descendants and which may, some say, have played a part in sparking off the Russian Revolution.

Haemophilia is a failure of blood clotting. This X-linked recessive condition already has a certain status as one of the few human mutations whose origin has been firmly established at a distance of several generations. Many of the male descendants of Queen Victoria (most notoriously the last Russian Tsarevitch, Alexis, killed in the aftermath of the Revolution) suffered from it, and many of her female descendants were carriers of the gene. Part of the royal haemophilia pedigree is shown in Figure 4.10 – it may be worth noting that the allele has been lost from the family tree branch leading to the present British royal family.

The symptoms appear, at first, straightforward. After a cut, the blood fails to clot and there is severe bleeding. Even worse, blood may leak into joints causing pain and, should it exude into the muscles themselves, muscle wasting. Other symptoms involve the formation of blood-filled blisters under the skin.

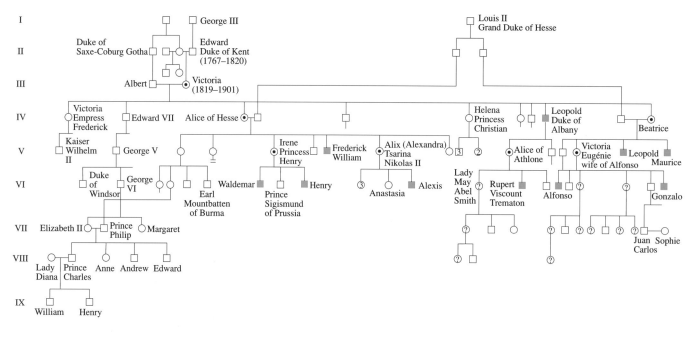

Figure 4.10 The inheritance of the X-linked recessive condition haemophilia in the royal families of Europe. A recessive allele causing haemophilia arose in the reproductive cells of Queen Victoria, or one of her parents, through mutation. This haemophilia allele spread into other aristocratic families by intermarriage.

⊙ carrier female   ▢ ⑦ status unknown

▪ haemophiliac male

Leakage of blood into the brain can cause brain damage or even sudden death. Until quite recently, there was no effective treatment for haemophilia, but now many of the symptoms can be cured by injecting an appropriate clotting factor.

The history of the understanding of the inherited failure of blood clotting follows that of many inborn conditions. Puzzling differences in symptoms or severity were first traced to differences in the site of the damage in a complex chain of biochemical reactions each involving a separate enzyme; in this case the cascade of reactions leading from cut to clot. Then it was found that a single step in the reaction can go wrong in different ways as one or another amino acid in a particular enzyme is changed. Finally, it became clear that the same kind of damage can arise in many different ways, most of which produce identical symptoms. This clinical 'learning curve' was accompanied by a transformation in geneticists' views of what the mutation is and how it takes place.

There are many steps between a cut and the formation of a clot and mutation can interfere in most of them. Some are so damaging that they kill their carriers, but others allow them to survive with varying degrees of disability.

A cut sparks off a long series of biochemical reactions, each under the control of a separate gene locus. The proteins involved (most of which are made in the liver) are circulating in the bloodstream ready for use, but in an inactive form. Activating them involves several biochemical steps and, for some, the involvement of vitamin K. Several of the genes are members of a gene family.

An injury stimulates **platelets** (fragmented blood cells without nuclei) to stick to each other with the aid of a specific protein to form an emergency plug. The platelets release chemicals that cause the vessel to contract. A clot is then formed. Its formation involves approximately 20 distinct steps (see Figure 4.11). The basic clot is a network of **fibrin** chains made from the soluble protein **fibrinogen** when it is cut by a circulating enzyme called **thrombin** (itself needing an enzymatic activator to set it into action). It is toughened up by the formation of cross links, catalysed by yet another enzyme that is again usually present in its inactive form. Other factors are also involved. One of these – **Factor IX** – is itself an activator of the so-called **Stuart factor**, responsible for the hardening of the clot. **Factor VIII** is an accessory in this process.

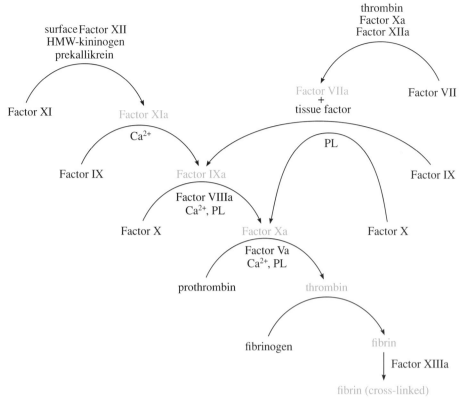

Figure 4.11 Abbreviated scheme for clot formation. PL = phospholipid; HMW kininogen = high molecular weight kininogen. The Factor VIIa-tissue factor complex is capable of transforming Factor IX and/or Factor X.

Until 1947 the existence of distinct forms of haemophilia was not known. In that year, though, it was found that blood from certain haemophilia patients could, when added to that from other patients, cause it to clot normally.

▷ What does this imply about the number of gene loci involved?

▶ There must be at least two. The two groups of patients differ in terms of which gene is damaged and so blood from one group can compensate for the absence of a crucial factor in blood from the other group.

The discovery that two clotting factors – two distinct gene loci – were involved was the first step in unravelling the complicated mutations that underlie the haemophilias.

Two of the genes involved in the final stages of formation of a fully functional clot are the cause of most haemophilias (and are in fact those discovered in 1947). They involve two blood factors, Factor VIII and Factor IX. One gives rise to the disease known as haemophilia A and is due to a mutation in Factor VIII. The other produces haemophilia B disease (which is sometimes known, after the first patient described, as Christmas disease) and is due to a mutation in Factor IX. Both are errors in blood clotting, but the effects of the A form of the disease are usually more severe than those of the B form. The A form of the disease affects around one in 10 000 males, the B form only a fifth as many. Both genes have now been cloned and sequenced.

Mutations have been found in many of the genes involved in the reaction cascade. However, most are extremely rare. There are, for example, fewer than fifty reported cases of inherited prothrombin deficiency. As is true for other mutations of severe effect, any error that interferes with a crucial step early in the clotting chain has such drastic effects on its carriers that they do not survive.

The Factor VIII gene spans 186 kb of DNA – about a thousandth of the X chromosome. It is chopped up into 26 exons and 25 introns and produces a polypeptide of around 2 500 amino acids (see Figure 4.12).

Figure 4.12 Schematic representation of the Factor VIII gene. Numbers of kilobases from the first exon of the gene are shown on top. Each exon is represented by a vertical line or filled box. Exons are numbered from 1 to 26.

The gene has undergone a great variety of mutations. Some have identical phenotypic effects, but the presence of a range of different DNA lesions can now be seen to explain the wide range of severity of the disease, from inconvenient to lethal. Figure 4.13 shows a few of the section deletions for the Factor VIII gene. Some involve just 2 kb of DNA, others the whole gene.

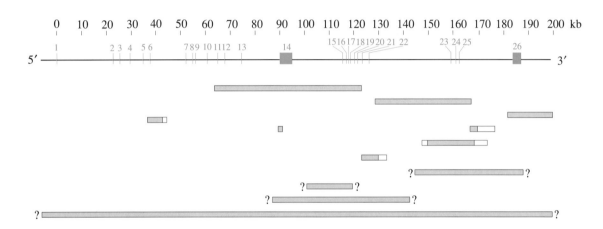

Figure 4.13 Some deletions of the Factor VIII gene in haemophilia A. The Factor VIII gene is shown on top. Each horizontal bar represents a different deletion within the Factor VIII gene. Open bars at the ends of some deletions denote the uncertainty of their extent. A question mark shows that the size of the deletion is unknown.

As well as mutation by deletion (which often causes severe symptoms) there is a variety of changes that involve just single nucleotides. Figure 4.14 shows a sample of these. Sometimes a single DNA change can produce a 'stop' codon and terminate the growing chain, resulting in a severe illness. Sometimes (as in a mutation which involves a change from CGA to CAA in exon 26 of the gene) there is a single amino acid change (from arginine to glutamine in this case) which does little damage to the molecule. Figure 4.13 shows that these single-nucleotide changes are clustered into one region of the gene. This is rich in CpG dinucleotides and is a mutational hot spot.

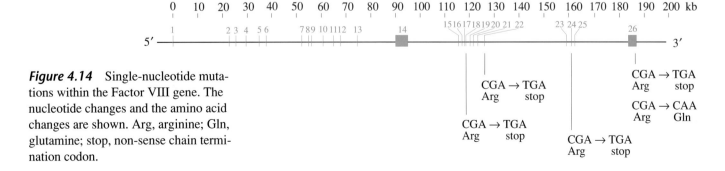

Figure 4.14 Single-nucleotide mutations within the Factor VIII gene. The nucleotide changes and the amino acid changes are shown. Arg, arginine; Gln, glutamine; stop, non-sense chain termination codon.

130

The Factor IX gene undergoes a similar diversity of mutational changes, although the gene is smaller than that producing Factor VIII. It is only 34 kb long and is divided into seven introns and eight exons coding for something over 400 amino acids. Figure 4.15 shows the structure of the gene and Figure 4.16 shows some of the deletion mutants for the haemophilia B condition. In addition, there is a range of mutations causing single base substitutions. As is the case for haemophilia A, almost every family with the disease carries a different mutation: the number is rapidly approaching 1 000 distinct mutations, world-wide. Compare this with the situation in say, sickle cell anaemia, in which a single mutation is shared by millions of people

Figure 4.15 Schematic representation of the Factor IX gene. A kilobase scale is shown on top. The gene has 8 exons (represented as filled bars).

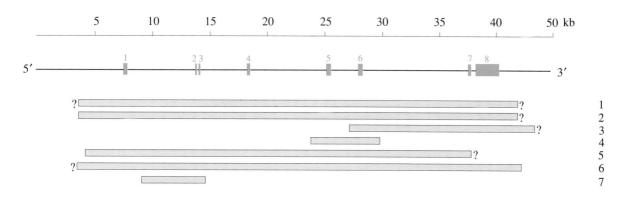

Figure 4.16 Deletions of the Factor IX gene in haemophilia B. Each horizontal bar represents a different deletion. A question mark denotes that the extent of the deletion is unknown.

Table 4.1 summarizes the nature of the mutation in a sample of 156 Swedish and British families known to carry haemophilia B.

Table 4.1 The incidence of mutation types within a sample of 156 haemophilia B families.

Mutation type	Incidence (number of families)
promoter	1
gross deletion	7
splice site	8
new splice site	5
frameshift	16
non-sense	23
mis-sense or amino acid deletion	96

There is a preponderance of mutations that alter – or delete – single amino acids. Those that damage crucial parts of the gene (such as the promoter or the splice sites involved in the editing out of introns, see Chapter 3) are much less frequent.

▷ Why is there such a variety of different mutations producing haemophilia – and why are single amino acid changes the commonest?

▶ As the gene is sex-linked, all males manifest its effects and, as it is certainly disadvantageous, each new mutation is likely to disappear within a few generations. Most haemophilia mutations are hence rather recent and confined to single, or a few, families, where they reside for a short time before they depart. Single amino acid changes are likely to cause less damage than, for example, the destruction of a splice site and can persist for longer.

As new haemophilia mutations are often identified, it is possible to work out the probable rate of mutation at individual sites in the gene. For the hot spot (the CpG islands) in the Factor IX gene, for example, this is about one chance in ten million of a transition per gamete per generation.

This heterogeneity of mutation has practical implications. For example, one of the complications of treating haemophilia with artificial Factor VIII or Factor IX is that some patients recognize the protein as foreign and produce antibodies against it. It transpires that these individuals carry mutations – such as large deletions or stop codons – that completely suppress the production of their own Factor VIII or IX protein. They produce an immediate immune response against the injected protein, and may die as a result. Now that the nature of every sufferer's mutation should be able to be worked out, patients unsuitable for treatment in this way can be identified before they are put at risk.

In addition, if every family has undergone its own more or less unique genetic accident, genetic counselling is made more difficult as each may need its own diagnostic test. However, technology is advancing so quickly that there is progress towards sequencing the gene itself in every affected family and identifying the lesion involved. In this way a national data-base of the precise mutational changes in every haemophilia A and haemophilia B family can be drawn up and used to give genetic advice to those at risk.

The message that emerges from the haemophilias is that what seemed to be a simple and unitary genetic change blocking the action of one of the body's important functions – the clotting of the blood – represents a great diversity of mutations, involving several different loci and hundreds of distinct mutational changes within them. The same picture, no doubt, will emerge as other genes are studied in the same detail.

4.7 Artificial mutagenesis – increasing the natural rate of mutation

The rate of mutation can be changed by artificial means. As the old image of mutation was a series of breakdowns in the genetic machinery, this discovery initially produced concern that the benefits of civilization, such as radiation and toxic chemicals, would increase the rate of mutation and hence the '**genetic load**' – the burden of hidden genetic disease which every human population carries. H. J. Muller, who discovered artificial mutagenesis, knew that many

Drosophila mutants were lethal and that others were harmful. He became very concerned that the same was true of human mutations and he wrote an odd dystopian novel – *Out of the Night* – predicting that doom would result from damage to the genetic material.

Since Muller's day, molecular biology has shown that mutation is common, and that many new mutations have no discernible effects on fitness. It is also clear that the human gene pool is not – as Muller thought – a pure and unpolluted reserve of perfectly functioning genes containing one or two unfortunate errors, but that diversity is universal. However, new insights into the relationship between DNA damage and the uncontrolled proliferation of body cells which gives rise to cancer have renewed the anxiety about **mutagens**, the external agents that cause mutations. This relationship is explored in Chapter 7.

Muller himself studied the effects of X-rays. Figure 4.17 shows the range of the electromagnetic spectrum. Only a small part has any effect on the genetic material. X-rays (and to a lesser extent gamma-rays) are **ionizing**: they act on the molecules – most notably water molecules – surrounding the genetic material, to produce **free radicals**, bearing an unpaired electron. This turns them into highly reactive chemicals which interact with any neighbouring molecule – including DNA. The damage may involve actual breakage of the chain or chemical alteration to one of the bases.

gamma-rays	X-rays	ultraviolet	infrared	microwave	radio
10^{-3}	10^{-1}	10	10^3	10^5	10^7	10^9

visible | | | | | wavelength/nm

Figure 4.17 The spectrum of electromagnetic waves ranging from radio waves to gamma-rays.

Ultraviolet rays can also damage DNA, but are much less penetrating than X-rays. They, too, do much of their damage by chemical means, in this case by causing adjacent thymine bases to become covalently linked to each other (dimerization. – see Figure 4.18).

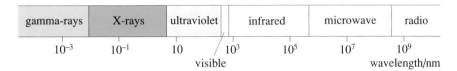

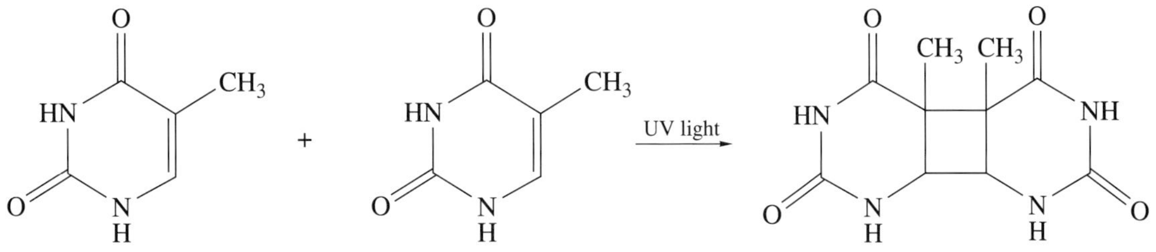

thymine   +   thymine $\xrightarrow{\text{UV light}}$ thymine dimer

Figure 4.18 Dimerization of two adjacent thymine molecules in the same DNA strand by the action of ultraviolet light. This thymine dimer produces a distortion in the DNA helix and disrupts the hydrogen bonds between the thymines and the adenines in the opposite DNA strand.

Muller worked on the induction of sex-linked lethals in *Drosophila*, using the approach already described (Section 4.2.1) He irradiated males with various doses of X-rays before mating them with virgin females of the special chromosome stock. The absence of wild-type males in the F_2 generation of such a cross showed that a mutation had been induced in the sperm of the original male.

Figure 4.19 shows the simple relation between X-ray dose and the incidence of recessive lethal mutations. Without radiation, there was a rate of one or two lethal mutations per thousand chromosomes tested. A dose close to that which kills or sterilizes the flies can multiply the number of mutations by a hundred times, but even a tiny increase in radiation increases the mutation rate.

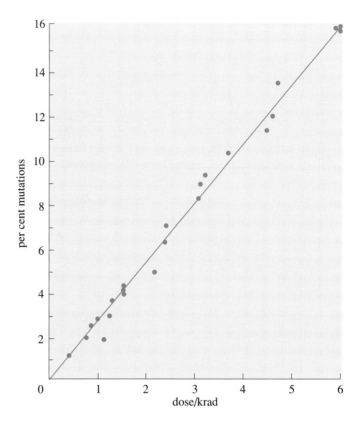

Figure 4.19 Linear relationship between X-ray dose to which *Drosophila melanogaster* are exposed and the percentage of mutations (mainly sex-linked recessive lethals). The units are expressed as krads, one rad being the radiation absorbed dose or $10^{-2}\,\mathrm{J\,kg^{-1}}$.

The linear relationship between dose and effect suggested to Muller that the gene could be seen as a target: the more bullets (in the form of X-rays) that were shot at it, the more likely it was to be hit.

▷ If the target theory is correct, what does this say about the dangers of low doses of X-rays (such as those used in medicine)?

▶ In principle, a small dose of radiation over a long time should produce the same number of mutations as a large dose given all at once.

No-one enjoys the prospect of being shot at, even if only a few bullets are aimed in one's own direction: after all, one lucky hit and the damage is done. This, thought Muller, might have alarming implications. Even a tiny long-term increase in radiation (resulting from medical X-rays, or atomic testing) would in time produce the same number of mutations as would a sudden nuclear disaster. Fortunately, though, experiments on mice show that a chronic low dose of radiation is much less effective in producing mutations than is the same dose given in one burst. A series of **repair mechanisms** acts to rectify low levels of mutational damage as it happens.

Another remarkable aspect of X-ray mutagenesis is that such a small amount of energy is effective in producing lethal mutations. This was the first hint that the genetic information is contained in a tiny part of the cell – a part which we now know, of course, to be DNA.

Muller's work led to great concern about the possible effects of radiation on human genes. The atom bombs were dropped on Hiroshima and Nagasaki in August 1945. Within a few months a programme of genetic research began. Some biologists feared that the offspring of those who had survived the bombing were likely to suffer from a wave of new genetic disease. Mothers were persuaded to take part in the research by giving them special ration books in exchange for their cooperation.

For the first few years, only very simple genetic tests were available; observation of gross deformities and slowing of growth, or changes in chromosome number. Later, two-dimensional protein electrophoresis (Section 4.3) was used to compare parents and offspring. Several million gene loci were looked at. The result was straightforward. Only four new mutations were detected in the offspring of parents exposed to the blast compared with two in a control group of children whose parents had been outside the city at the time. There was no evidence, from this study at least, that the people of Hiroshima and Nagasaki – the most heavily irradiated population in history – had suffered any significant increase in non-lethal mutation rate.

This discovery has been used by some to suggest that radiation safety standards are too stringent, although others dispute any suggestion that it would be safe to relax them. The story is complicated by uncertainties about just how much radiation was produced by the bombs and, of course, could only look at mutation rates in live births. Furthermore, the link between radiation, somatic mutation and cancer (see Chapter 7) means that the effects of radiation on human genes is still a matter of considerable interest.

▷ Is it safe yet to assume that the Japanese bombings produced no genetic damage?

▶ Certainly not: in spite of the large numbers of genes investigated it is possible that some mutations were missed; and, more important, any new recessive mutations in functional genes might not manifest their effects for many generations.

The link between mutation and military research goes further than the Hiroshima and Nagasaki experiences. In the late 1930s the young geneticist Charlotte Auerbach moved from Germany to Edinburgh. She was intrigued by the observation that certain chemicals (notably the war gases, such as mustard gas) produced painful burns that were similar to those caused by an intense burst of radiation and which took a long time to heal. Soon she found that the chemicals involved could cause mutations. Her discovery was kept secret until the end of the war. Since then it has given rise to the study of **chemical mutagenesis**.

Some substances have a dramatic effect on the mutation rate. For example, one chemical, acridine mustard (a relative of mustard gas) increases it by 5 000 times when tested on the fungus *Neurospora*. Few substances are as potent as this, but there are many that are more effective at producing mutations than is the highest non-lethal dose of X-irradiation.

A bewildering variety of chemicals cause mutations. Some, the **base analogues**, are related to the DNA bases themselves, and are incorporated into the DNA chain as it is synthesized. Others react with natural bases, altering their pairing ability and generating mutations. Ethyl methyl sulphonate, an **alkylating agent**, damages DNA, in this case by adding ethyl groups. More simple chemicals such as nitrous acid remove the amino group from cytosine, transforming it into uracil.

Other chemical mutagens act more subtly. Some, such as acridine orange, are not closely related to the DNA bases, but have approximately the same shape and can slip into the replicating chain. These small insertions cause "frameshift" mutations, that is, they alter the position of the reading frame (see Chapter 2).

Some chemicals, the **DNA adducts,** bind to DNA, causing damage as they do so. Aflatoxin – a substance common in the moulds that grow on badly stored food – does this, as do the benzopyrenes found in tobacco smoke and petrol fumes.

Such binding of DNA adducts may be a first sign of mutational damage – either to germ cells or to somatic cells which might become cancerous as a result. There is now known to be a close relationship between a substance's capacity to cause mutational damage in bacteria and its ability to cause cancer. The Polish city of Gliwice has some of the worst air pollution in the world, largely because of the wide use of soft coal for heating. The extent of adduct binding goes up in the winter, giving rise to alarm about DNA damage.

4.8   DNA repair: patching up the damage

DNA is susceptible to damage both by external agents and through its own intrinsic instability. Three thousand million years ago, when life was little more than a series of nucleic acid chains, the rate of mutation must have been very high. Since then there has evolved a series of effective mechanisms which reduce DNA damage, either by destroying dangerous substances as they appear, or by restoring mutated DNA to its original form. Without them, life would not be possible.

As mentioned earlier, cytosine is often converted spontaneously to the related base uracil (whose DNA pairing preferences are similar to those of thymine). This natural rate of mutation is so high that perhaps one in 1 000 cytosines is expected to undergo chemical degradation in a single human generation – which would certainly be lethal. The repair mechanisms ensure that the damage is greatly reduced.

Cells contain a variety of enzymes that break down harmful chemicals before they have a chance to attack DNA. **Superoxide dismutase,** for example, destroys highly reactive oxygen radicals as they appear. It produces hydrogen peroxide, which is in turn broken down by **catalase. Cytochrome oxidase** is a related system. Genetic variation in the activity of such enzymes plays a part in individual susceptibility to carcinogenic chemicals.

There is an array of mechanisms for repairing the DNA, should this first line of defence fail. Certain enzymes can reverse chemical changes to individual bases. Others cut out the length of DNA which contains a mutation, and synthesize a new section. Such **excision repair** (see Figure 4.20) needs several components; an endonuclease to make the cuts, a polymerase to synthesize the replacement chain, and a **ligase** to stitch it into the gap.

Some mutations slip through even this net. Several second-line defence mechanisms are available. **Mismatch repair** depends on the recognition (manifest in the distortion of the double helix) of mispaired bases after DNA replication. A series of enzymes cuts out and replaces the inappropriate base.

There is a hierarchy of additional mechanisms in both bacterial and eukaryotic cells, some of which can even reconstruct DNA when the information from both strands has been lost through mutation. They do this by 'referring to' other copies of that DNA which may be present in the cell (for example, in the multiple replicas of the circular chromosome often present in bacteria).

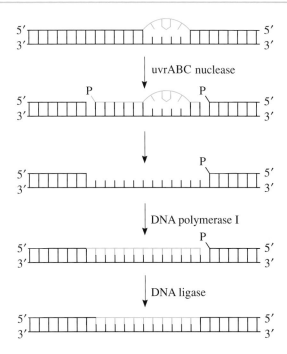

Figure 4.20 A model for repair of mutation in a bacterium. Here uvrACB nuclease cuts out a 12-nucleotide-long single-stranded DNA fragment. The section carrying the damage is removed, and the gap is filled by DNA polymerase I (which synthesizes a new and complete strand of DNA) and sealed by DNA ligase.

However they work, the systems for reversing mutations are efficient. In bacteria, only about one base in 10 000 million is altered each generation. In strains whose repair mechanisms have themselves been damaged by mutations, this rate may rise by 1 000 times.

Most of the repair machinery depends on the fact that the information in DNA is held in double copy, one in each strand of the helix. The undamaged strand is used to restore the one struck by mutation. Perhaps this is why genomes based on a single strand of DNA – and so comprising a single copy of the genetic information – are never more than about 5 000 bases long and are restricted to intracellular parasites such as phages. For more complex genomes to evolve there must be a 'back-up' of the genetic information which can be drawn upon when repairing damage.

Repair enzymes play a similarly crucial role in humans. People with the inborn disease xeroderma pigmentosum are very susceptible to sunlight. Even a small dose of ultraviolet leads to the development of dark freckles, many of which develop into skin cancers. Most of the individuals affected die before they are 30 years of age. Their metabolic defect lies in one of the enzymes involved in excision repair.

4.9 Sex, age and mutation

Meiosis in males never ceases as sperm are produced constantly throughout a mammal's life. In many creatures, though, eggs enter the process of gamete formation just once in a female's existence. In most mammals egg development is frozen early in meiosis as the female becomes sexually mature. They are released at intervals during her reproductive lifetime.

This means that there are more cell divisions between each generation's formation of sperm than there are for production of eggs. Every division incurs the risk of a mistake, so that there are more chances for mutations to happen in the male germ-line than in the female. The rate of mutation is higher in males than in females, so much so that most new mutations for inborn disease happen in the paternal line.

The strongest evidence comes from the study of new mutations for haemophilia. A haemophiliac son may appear in a family either because the gene has been present for some time or as a result of a new mutation. That mutation might have occurred on one of the maternal X chromosomes (not, of course, that of his father, which does not descend to sons), or an X chromosome descending from a grandparent.

By examining the length of DNA around the actual mutation site (the haplotype; Chapter 1), it is possible to determine the X chromosome on which it actually took place. Sometimes, the affected son's haemophilia is carried by both his mother and one of the two grandmothers: it descends from an earlier generation and its origin usually cannot be traced. In other cases, though, the mutation happened in the recent past. Occasionally, the patient's mother does not carry the mutation in her body cells: it took place as her egg was formed. More often, she is a carrier, but neither of her parents is. The genetic accident giving rise to haemophilia must have happened when one of her own X chromosomes was formed. Comparing the DNA sequence around the damaged gene with that in the appropriate section of her father's and mother's X chromosomes tells us which of her parents was responsible for the mutation.

By comparing patients, their mothers and their grandparents it becomes possible to work out in whose cells – mother's, grandmother's or grandfather's – the mutation took place. An extensive survey of this kind has been undertaken on haemophilia B patients in Sweden. The result is clear: the chance of a new haemophilia mutation taking place as a sperm is formed is ten times greater than that in an egg. Mutation is, perhaps, mainly a male concern.

▷ If there is a higher rate of mutation in males than in females, what do you expect – other things being equal – the relative rate of autosomal and sex-linked mutations to be?

▶ Sex-linked genes spend relatively more of their time, in evolutionary terms, in females than in males: a particular X chromosome has about two-thirds of its evolutionary history within a female and – because of the existence of a Y chromosome – only one third in a male. Autosomes, of course, have an equal probability of passing down through male or female lines. We would hence expect X-linked genes to accumulate fewer mutations than autosomal. Comparison of related species of a variety of creatures, from *Drosophila* to mice, shows that (at least for mutations which have only small effects on their carriers, such as those in introns) there has indeed been a slower rate of mutation, as measured by overall divergence, for X-linked than for autosomal genes. The difference suggests that the overall mutation rate in males, summed over generations, is several times that in females.

The longer the generation time, the greater the number of cell divisions within males before the crucial successful sperm is produced. Mice and rats mate when the male is a few months old. Given the rate at which sperm and eggs are produced this means that the number of cell divisions separating a father's sperm from that of his offspring is about twice that separating the eggs of mother and daughter. Humans wait until the male is about 20 years old before reproducing, so that there are many more times as many mitoses (each with a chance of an error) per generation in the male than in the female germ line. One estimate is that the ratio of mutation rate in male compared to female primates is ten to one, while in mice it is only two to one.

Another important mutagen is simple, important and unavoidable. It is old age. Elderly parents are much more likely to produce genetically damaged sperm or egg than are those less advanced in years. Figure 4.21 shows the rate of mutation with increasing parental age compared with the rate in cancer patients undergoing radiotherapy.

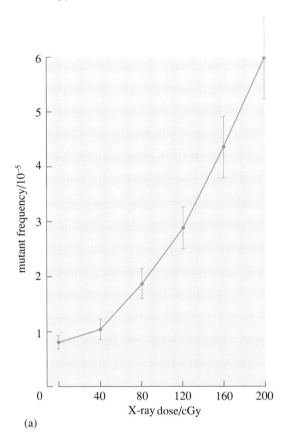

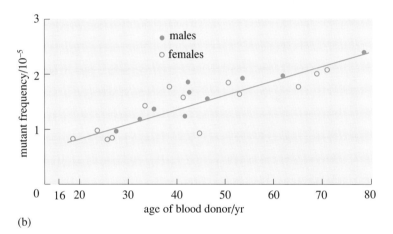

(a)

(b)

The effect of age is dramatic and, because of its inevitability, is likely to outweigh the influence of any external mutagen likely to be experienced. It is largely due to the decrease in effectiveness of repair enzymes as old age advances as well as to the increased numbers of cell divisions experienced by the cells producing gametes. The effect of age on mutation is particularly striking in somatic cells, and explains in part the large increase in cancer incidence among the very elderly (Chapter 7).

Figure 4.21 Relationship between the incidence of new mutations removing the activity of an enzyme in white blood cells of blood donors (a) exposed to X-rays during cancer therapy (dose units in Grays), and (b) of increasing age.

4.10   Has the rate of mutation evolved?

Table 4.2 shows the mutation rate from a series of loci in a variety of creatures. It can, it seems, differ in rate by 100 000 times. Why should this be – and why, indeed, is the frequency of mutation confined to this range?

The rate of mutation differs greatly between loci and can certainly evolve. The existence of repair mechanisms suggests that the rate has been greatly reduced by natural selection since the first genes appeared. It can also be altered experimentally. Populations of *Drosophila* exposed to low levels of radiation in the laboratory for many years are more resistant to the mutagenic effects of a sudden high radiation dose than are others.

Table 4.2 Spontaneous mutation rates of specific genes.

Species and locus	Mutations per 100 000 cells or gametes
Escherichia coli (bacterium)	
streptomycin resistance	0.000 04
resistance to T1 phage	0.003
arginine independence	0.000 4
Salmonella typhimurium (bacterium)	
tryptophan independence	0.005
Neurospora crassa (fungus)	
adenine independence	0.000 8–0.029
Drosophila melanogaster (fruit-fly)	
yellow body	12
brown eyes	3
eyeless	6
Zea maïs (maize)	
sugary seed	0.24
I to *i*	10.60
Homo sapiens (human)	
retinoblastoma	1.2–2.3
achondroplasia	4.2–14.3
Huntington's disease	0.5
Mus musculus (house mouse)	
a (coat colour)	7.1
c (coat colour)	0.97
d (coat colour)	1.92
In (coat colour)	1.51

In some cases, too, the rate of mutation can determine a population's ability to respond to a change in the environment. Bacterial cultures containing a gene that produces a general increase in the mutation rate evolve new biochemical pathways more quickly than do controls without the gene when each is exposed to a new synthetic food source. The fact that the mutator genes (see Section 4.5) are not common in nature suggests that, in the real world, bacteria rarely face such unexpected demands.

Bacteria provide another hint about the evolution of the mutation rate. Pathogenic bacteria must adapt rapidly to their host's defences against infection. Often, the frequency of a particular cell-surface antigen on the bacterium shifts greatly during the course of an infection, with one new antigen succeeding another. *Haemophilus influenzae* (which causes meningitis) attacks its host in this way, using a complex set of distinct antigens to do so. It can also produce *fimbriae*, protein 'hooks' that allow it to bind on to epithelial cells where required. The genes for each of these systems have a very high mutation rate. Many of the mutations produced (some of which are advantageous and spread rapidly through the bacterial population) arise because of inaccurate pairing between repeated sequences.

▷ Can you recall an example of a human disease where a similar mechanism involving repeat sequence mutation is at work?

▶ Huntington's disease, where the number of copies of a CAG repeat increases as it passes from father to child.

Perhaps there is an optimal mutation rate for most genes. Some, filled with repeated sequences which reduce their stability, permit evolutionary flexibility in the face of an unpredictable environment. Others – the '**housekeeping**' genes (Section 1.5.3) – carry out important and unchanging tasks within the cell, and have a structure that produces far fewer new mutations.

4.11   Envoi: mutation as a natural property of the genome

Mutation was once seen as a process analogous to hitting a watch with a hammer: a disruption to a finely tuned mechanism that was almost certain to cause damage. Now we know that mutation is intrinsic to the genetic machinery itself. It is involved not only in evolutionary progress, but in cancer and old age. This balance between the two is considered at greater length in Chapter 7. Once, mutation was just seen as a mine of raw material from which evolution can fashion new products. The new vision of the genome as a more fluid system constantly under attack from mutation has, inevitably, altered this view. There is a new acceptance that mutation has a dynamism of its own that can drive the evolutionary fate of segments of the genetic material. The new technology of genetics has changed our views of many of its processes – mutation, perhaps, most of all.

Objectives for Chapter 4

After completing this chapter, you should be able to:

4.1   Define and use, or recognize definitions and applications of, each of the terms printed in **bold** in the text.

4.2   Discuss changes in our view of mutation as technology has advanced and understand how mutations are detected, both indirectly (as in Muller's experiments with *Drosophila*) and directly by protein electrophoresis or DNA analysis.

4.3   Interpret patterns of chromosome mutation; changes in ploidy, deletions, duplications, inversions and translocations.

4.4   Discuss the nature of mutational change at the DNA level, including the importance of transposable elements in inducing mutations and the new concept of the fluid genome.

4.5   Integrate this information using a study of mutations in the blood-clotting system; why mutations in most of the loci involved are almost never found, how haemophilia A and haemophilia B arise, and the diversity of changes that can damage each one of them.

4.6   Assess the relative importance of various mutagens including radiation, chemicals and age.

4.7   Discuss the various mechanisms involved in the repair of mutations, ranging from enzymes that break down oxygen radicals to the complex pathways of DNA repair.

4.8   Place the process of mutation into a biological context and discuss how it may have evolved, and how there are differences in the rate of mutation between the sexes.

Questions for Chapter 4

Question 4.1   (Objective 4.2)

How have changes in the way that mutations are detected altered views about their nature?

Question 4.2   (Objectives 4.3 and 4.4)

Compare mutational events at the chromosomal and the DNA level.

Question 4.3   (Objective 4.6)

What changes in the human mutation rate are likely to arise from the benefits of civilization?

Genetics and evolution

5.1   Introduction – Darwin, Mendel and a synthesis

The theory of evolution – of descent with modification, as Darwin called it – turns biology into a science rather than a specialized kind of stamp collecting. It permeates all inquiries into the living world. Evolution contains genetics, the machinery of descent, within itself and makes sense of how genes work and what they are. When it comes to the sorts of characters that Darwin himself was concerned with – variation in visible characters, or organs such as legs or tails – his model of evolution is in robust health. Even the evidence from the fossil record and from comparative anatomy, both crucial to his argument in *The Origin of Species,* has been strengthened by modern genetics as each is, in general, strongly supported by the trees of relatedness which emerge from studies of DNA. Nobody doubts the central role of natural selection – inherited differences in the chances of reproduction – in generating the adaptive changes which allow each organism to survive in the struggle for existence. Many of them are discussed in the Open University course S365 *Evolution: a Biological and Palaeontological Approach.*

However, genetics has outgrown its Darwinian parent. The structure of genomes is so startling that new evolutionary processes, many of which would have astonished Darwin, are needed to understand what is going on. This chapter discusses how evolutionists have succeeded – and failed – in understanding evolution now that the physical map of the genes is almost complete. It shows how intimately the study of organisms and of genes are related; and how genetics makes sense of Darwin's theories. Many of the findings of molecular biology fit into the argument of *The Origin of Species*: that evolution happens through mutation, natural selection and accidental events such as the colonization of new islands by small groups of founders. We saw in the previous chapter how much the image of mutation has changed with developments in molecular technology. The theory of evolution has itself been forced to adapt to these startling new findings.

However, molecular biology can support Darwin. Descent with modification is proved by looking at the genes of evolving lineages. Some change quickly: viruses show how, within a few years, one genome can evolve into another. Even when the split between two lines of descent took place long ago, comparing the genes of living creatures (humans included) proves their common ancestry. Given some daring assumptions about the rate of evolution, DNA sequences can also be used to date when two lineages separated. Natural selection, the engine of evolution, can produce molecules with new properties; and the course of selection in the wild retraced by looking at patterns of DNA sequence.

But all is not sweetness and light for Darwinians. The discovery of huge amounts of variation places more emphasis on the role of accident than Darwin might have approved of. Even worse, much of the structure of the genome is repetition – the existence of sequences with no apparent function – and this together with the rapid divergence between closely related species demands new explanations. To think of genes, rather than those transmitting them, as the targets of selection may help: but Darwinism can provide only a partial explana- tion of molecular evolution. Genetics is nevertheless crucial to the modern theory of evolution as it confirms Darwinism's central dogma: that all organisms are relatives, and that all descend from shared ancestors that lived long ago.

The mechanism of evolutionary change put forward in *The Origin of Species* in 1859 is simple: as Thomas Henry Huxley, Darwin's great defender, said, 'How very foolish not to have thought of that!' It is based on inherited differences in the chances of reproduction.

The theory depends first on the fact that the transmission of information between generations is imperfect; there is *mutation*. As a result, a slightly altered version of the message is passed on each time. This, the first part of the Darwinian machine, generates inherited diversity.

Its second part depends on the almost universal biological fact that organisms produce more offspring than their environment can support: inevitably, not all can survive to pass on their own genes. There is, in other words, a **struggle for existence**. A crucial part of the Darwinian mechanism is **natural selection**: this is a filter which rejects certain variants and favours others as they fight the battle for genetic survival. Sometimes, a mutated gene interferes with its carrier's ability to survive and reproduce and it disappears, either immediately or after a few generations. Occasionally, though, a new variant is better at coping with what life throws at it than what went before. Its carriers thrive, breeding more effectively than do their relatives. As a result, the new variant becomes more common in subsequent generations.

Darwin believed that such a 'preservation of favoured races in the struggle for life' could, if it went on for long enough, explain how the creatures of today arose from their simpler predecessors of long ago.

A final part of his mechanism – which has gained importance in the light of modern genetics – involves evolution as a result of chance events. It might be, for example, that a few individuals – with their biological heritage – are blown to an ocean island, where they flourish. The accidents of sampling may mean that the emigrants do not contain within themselves all the diversity present in the original population: the island population has evolved as a result of a random occurrence, or – as it now known – by **genetic drift**.

Darwin felt that if his machinery of evolution were given enough time, then slowly new forms of life – new **species** – would evolve. Central to his theory was the idea that the minor evolutionary changes seen within existing species are enough to explain the major events of evolution, the origin of species included. No aspect of his work has produced more controversy than has this claim.

Darwin knew nothing of how inheritance actually worked. As you saw in Chapter 1, he had a scheme in which the attributes of parents were blended in their offspring. He soon realized that if inheritance was based on fluids – the mixing of bloods – his idea was in trouble. Any advantageous changes would be diluted out. In later life he paid much attention to this problem, without success.

Mendel provided the answer. Genes were particles. They could pass down a lineage unchanged and any useful innovations could be recovered in later generations.

▷ Which will spread more rapidly through natural selection – an advantageous recessive or an equally advantageous dominant allele?

▶ A dominant shows its effects in all its carriers. A recessive, though, is evident only in homozygotes – which, for new alleles, are extremely infrequent. Perhaps this explains why many (but not all) rapidly spreading mutations to new and suddenly advantageous characters like industrial melanism and insecticide resistance are dominant to wild-type.

For many years there seemed to be an inherent contradiction between Darwin's belief in slow, almost imperceptible change and the large (and usually damaging) mutational changes studied by geneticists. By the 1930s, though, a sophisticated mathematical structure of theoretical population genetics had accommodated genetics into evolutionary theory: the accumulation of many mutations of individually small effect in evolving lineages could, it seemed, explain most evolutionary patterns. This so-called 'modern synthesis' of evolution and genetics persisted successfully until quite recently.

Evolution began without genetics, but the subjects are now inseparable. How and when mutations take place, whether natural selection happens constantly or scarcely at all and whether species arise through gradual changes within existing lineages are at the centre of modern evolutionary theory and molecular biology.

5.2 Molecular genetics and evolution – a new schism?

Molecular biology has, though, led to an upheaval in the theory of evolution. Some of the turmoil can be resolved by making adjustments to traditional views. Darwin himself accepted that some evolutionary change took place at random. The study of DNA has produced a new awareness that many mutations are indeed neutral (Chapter 4) and have no effects on the survival and reproduction of their carriers. Their persistence or otherwise is determined by chance.

▷ Why is it hard to see how strong natural selection could act on all, or even most, DNA variants?

▶ Selection has a cost: every individual that fails its test dies or does not pass on its genes (think, for example, how expensive sickle cell haemoglobin is in terms of the loss of segregating SS homozygotes to the populations that possess it; Chapter 1). It is hard to see how any population could afford the losses involved if selection worked simultaneously, and independently, at millions of variable sites in the genome.

The ability to look directly at the genes of related species also shows that Darwin's view that the origin of species is just a continuation of natural selection by other means is oversimplified. The divergence of some species from their progenitors seems to involve large numbers of genes: for others, though, far fewer are involved.

None of this strikes at the heart of the classical theory of evolution. Such controversies are a matter of emphasis. However, it is beginning to look as if much of the structure of the genome may not be explicable in Darwinian terms at all. New evolutionary mechanisms are called for and much of this chapter is devoted to describing them.

For example, some DNA sequences have – or can be interpreted as having – their own evolutionary agenda, which may not be the same as that of the organisms carrying them. Genes can, metaphorically speaking, behave in their own interests, even if these conflict with those of the organism, their temporary home. What matters in evolution is the survival of lineages, not individuals.

Sometimes, DNA is itself so volatile that single species, under a veneer of morphological stability, are in constant upheaval at the molecular level, with parts of their genetic material in flux as one or another DNA sequence sweeps through. Indeed, many evolutionary processes within species may – in direct contradiction to Darwin's views – be quite different from those causing new species to originate. Although Darwinians are fighting a rearguard action, some molecular biologists claim that most evolution is based on these novel mechanisms.

It is much easier to describe how things are than to understand how they got there. Think of the many different accounts that can be given of, for instance, the origins of the conflict in former Yugoslavia. So far, there is no universal theory of history; and no general doctrine of evolutionary genetics. Because their science spends much of its time trying to infer what went on in the past, evolutionists are an unusually quarrelsome lot. In compensation, their subject is at that most alluring phase of science as it comes to terms with a flood of new and startling results. The *fact* of evolution is not in doubt, despite the controversy: what is in dispute is *how* it takes place.

5.3   Descent with modification proved

The only illustration in *The Origin of Species* is of a gnarled oak showing the relationships between the living animals who occupy its twigs and their descent from a common ancestor at its root. Darwin's evidence for his famous tree was not very substantial. It depended only on what he knew about shared structures among evolving groups. Much was based on general impressions of similarity, which was all that comparative anatomy could provide.

▷   Why are simple anatomical comparisons sometimes misleading when trying to make an evolutionary tree?

▶   Think, for example, of **convergent evolution**: the appearance of the same structure in distinct lineages – the fish's and the whale's tail, for example, or the wings of insects and birds.

Genetics gives the first chance to test the idea of descent with modification: that lineages change over time and that evolution can be tracked by looking at differences in genes. All genes are messages from the past: living fossils, either persisting for millions of years, or modified by mutation. Comparing related species allows one to guess at the genome of their long-dead common ancestors.

The best illustration of how genes can be used in this way comes from an organism that evolves so fast that we can see it changing. It is the virus responsible for AIDS, the acquired immune deficiency syndrome. The human immunodeficiency virus, HIV, is a retrovirus (Chapter 2), adapted to living within one of the key cells in the immune system, the helper T cells (Figure 5.1). For part of its life it is a single strand of RNA. After infecting a new cell, it uses the enzyme reverse transcriptase to synthesize DNA from its RNA template and to transform the next generation of viruses into lengths of double-stranded DNA.

The enzyme makes many mistakes during the copying process, lacking the repair mechanisms present when DNA is replicated (Chapter 4). The rate of mutation – and evolution – of parts of the HIV genome is hence very high. It may be as much as one in 10 000 per nucleic acid base each generation. As the virus is around 10 000 bases long, there is likely to be at least one mutation each time it replicates.

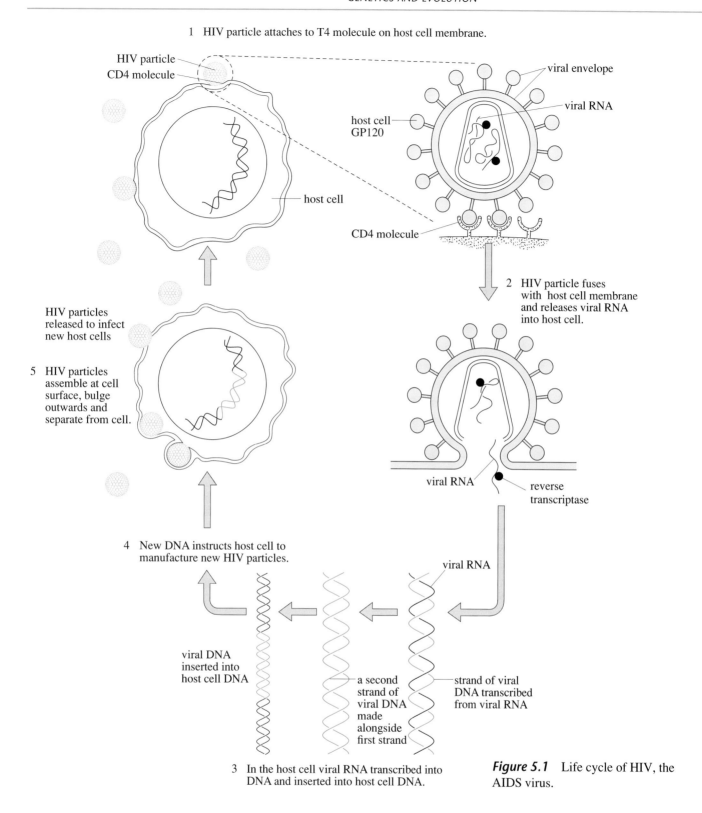

1 HIV particle attaches to T4 molecule on host cell membrane.

HIV particle
CD4 molecule

viral envelope

host cell
GP120

viral RNA

host cell

CD4 molecule

2 HIV particle fuses
with host cell membrane
and releases viral RNA
into host cell.

HIV particles
released to infect
new host cells

5 HIV particles
assemble at cell
surface, bulge
outwards and
separate from cell.

viral RNA

reverse
transcriptase

4 New DNA instructs host cell to
manufacture new HIV particles.

viral RNA

viral DNA
inserted into
host cell DNA

a second
strand of
viral DNA
made
alongside
first strand

strand of viral
DNA transcribed
from viral RNA

3 In the host cell viral RNA transcribed into
DNA and inserted into host cell DNA.

Figure 5.1 Life cycle of HIV, the
AIDS virus.

Someone with an advanced case of AIDS may contain 10 billion (10^{10}) virus particles. By the year 2 000, a hundred million (10^8) people may be infected with the disease. The evolving population of viruses will, by then, consist of (very roughly) 10^{18} genomes, each with a high rate of mutation.

All this makes AIDS a microcosm of evolution in action. Evolution in the virus is fundamentally not very different from that in, say, a mouse or a human being.

Some mutations may be favoured by selection although (as is the case for mammalian genes) we have no real idea which ones they are. Certain genotypes might, for example, be more infective than others: if so, this would be a powerful means of selection. However, the majority of new mutations probably do not differ greatly in their chance of surviving: they are neutral.

▷ What parts of a virus genome might be particularly open to the action of selection?

▶ Any gene altering its envelope protein (the site of the cell surface antigens recognized by its host) – and hence allowing the virus to keep ahead of its host's immune system – may be expected to evolve particularly quickly.

By choosing parts of the viral genome with the appropriate rate of change, it is possible to produce family trees which track the evolution of the virus within a patient, within a group of infected people and within the world's population of AIDS victims. It is even feasible, by comparing the virus with its relatives in other primates, to infer the source from which it began to attack human beings.

Very early in infection, most patients contain a single homogeneous population of viruses, suggesting that just a single virus is enough to initiate the disease. This may seem surprising given that thousands of blood cells are transmitted by, for example, the sharing of needles by addicts. It seems that the rate of evolution of viruses within a patient results in the majority soon losing their ability to infect, as crucial segments of the genome are deleted by mutation. Only a few stay complete enough to be infective.

Because of the high rate of mutation, this single founder in its new home soon evolves into a wave of genetically distinct invaders. The virus evolves so quickly that a single patient may contain viruses differing from each other in one-tenth of their base sequences. This is the key to the disease's virulence. The immune system is very good at clearing viruses from the body, but in AIDS it is soon overwhelmed by their number and by their diversity. Because of mutation of the virus, the immune system has to constantly face enemies never previously encountered. The patient dies from an immunological surrender which allows other pathogens to enter unchallenged or cancer to spread with no hope of respite.

As well as allowing the course of infection to be tracked through the increasing diversity of the viruses, the rapid evolution of HIV has practical implications. Vaccines work by persuading the body to produce antibodies against a specific antigen. If the virus keeps changing its identity then a vaccine will be hard to find. The high mutation rate also means that there is every prospect of the virus evolving drug resistance once treatment has started: this is now common among patients who respond well to drugs early in infection, but who die because the drug leaves some mutated viruses unscathed.

The evolution of the AIDS virus can be used to trace its history over longer timescales: in a population, rather than an individual. Every AIDS outbreak is a microcosm of the process of evolution. One group of patients were all clients of a Florida dentist, who continued to practise for two years after he was diagnosed as being infected with the virus in 1987. The DNA sequence of the 'envelope gene' (a rapidly evolving section of the genome) of viruses from seven infected individuals who had been his patients was tested. The virus from five differed in sequence by less than 5% from that of the dentist: that of the other two by 11% and 14% respectively.

A sample of other HIV-positive individuals from Florida who had never visited the dentist's surgery showed that their viruses differed by more than 10% from those of the dentist. Figure 5.2 shows the pattern of change in the dentist, his patients and in a sample of the general population of Florida ('local controls'), and the evolutionary tree that links them.

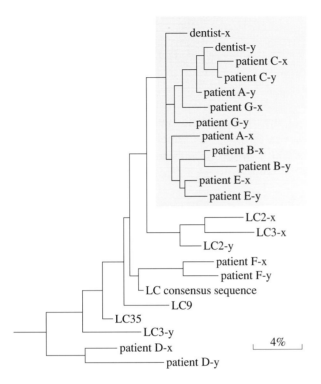

Figure 5.2 An evolutionary tree of a Florida population of AIDS viruses. The analysis compared envelope gene sequences from the HIV-positive dentist, his patients A–G and and some local controls, LC2, LC3, LC9 and LC35. LC consensus sequence denotes the control that is constructed from the most frequently occurring nucleotides at each position and hence represents an 'average' of the local controls.
'x' and 'y' denote the two most diverged sequences from each person. The group of sequences from the dentist and the sequences most closely related are shown in the shaded box.

▷ Patients D and F stand out from the others on this tree. Why might this be?

▶ They were probably infected from independent sources. In fact both these exceptional patients admitted to 'high risk' (homosexual or drug injecting) behaviour. The family tree of the AIDS virus suggests that their infection came from one of these activities rather than from the dentist.

This genetic evidence was crucial in settling the insurance claims against the dentist's estate made by the relatives of the patients who died: the unfortunates who contracted the virus in other ways got nothing. The genes of the viruses had been used to reconstruct their evolutionary history.

The AIDS evolutionary tree has some longer branches, too. By choosing sections of the genome that mutate at the appropriate rate it is possible to trace its ancient history. Although there is controversy about the details, a tree of relatedness of viruses from different parts of the world points at an African origin, with a passage first to the Caribbean and then to the United States. Within Africa, some of the strains nearest the root of the tree are in West Africa.

Some of these viruses, in the less promiscuous populations, produce less virulent symptoms. This is a hint that natural selection (based, in this instance, on the need to keep hosts alive until they have another sexual partner) is involved in the evolution of the AIDS virus. If the host is promiscuous the virus will be passed on soon after the first infection. It gains no advantage from moderating its own rate of replication and can afford to damage its carrier with no risk that he will die before infecting someone else. In places where carriers behave differently – perhaps with a second sexual partner only years after the date of infection – evolution favours virus genomes that have less harmful effects on the host (who can then stay alive for long enough to infect another victim).

▷ What other kinds of disease organism are likely to evolve changes in virulence associated with changes in human behaviour?

▶ One example is that of waterborne diseases. Cholera became more damaging after the introduction of the first water-closets (which pumped sewage into streams used for drinking water). Before then, a patient rarely passed the bacteria on. The flushing toilet meant that one patient could infect hundreds more; and needed to be kept alive as a source of infection for only a few days. Modern sewage treatment has, once again, reduced the infection rate, and selection has produced cholera strains that kill their victims less quickly.

The longest branches on the AIDS tree are those which connect the human virus to its relatives elsewhere in the animal kingdom. Many mammals – cats, mice, monkeys and others – have retroviruses not unlike that which causes AIDS. Those closest to the human virus are found in primates. One contender for the role of closest viral relative is found in the African green monkey. Perhaps this virus attacked humans as they hunted monkeys and were bitten by their injured prey.

The AIDS story shows how genetic change can be used to retrace the course of evolution. Darwin would have approved. The AIDS viruses of Florida dental patients or of old world monkeys fit perfectly onto his picture of an evolutionary tree. A central principle of *The Origin* has been confirmed by modern genetics.

5.4 Humans as animals: genes and our place in the living world

In *The Origin of Species*, Darwin dared say nothing more about human evolution than 'light may be cast on man and his origins'. In 1871 he went further, and in *The Descent of Man and Selection in Relation to Sex* he made a strong case that humans are related to apes and that humans and primates are descended from a common ancestor.

This case was made almost entirely on shared similarity: no fossils recognized as human were then known. The characters he used – skeletons, or behaviour – are certainly under the control of genes, but even now nothing is known of just how many genes might be involved. Modern genetics provides a compelling case that humans are indeed apes, and that the evolutionary rules that propel the AIDS virus are not very different from those leading to genetic change in more pretentious creatures. Genes mean that 'descent with modification' can be proved in retrospect as convincingly as it has been in organisms like the AIDS virus, in which the process takes place as a physician looks on.

Although biologists after Darwin accepted that humans were indeed apes, it seemed reasonable to suppose that they were remarkably special apes. To our eyes at least, *Homo sapiens* looks quite different from its relatives. Perhaps, then, its genes are equally distinct: humans and chimps may be cousins, but distant ones.

The idea of humans standing alone on a genetic pinnacle began to disappear when it became possible to look at individual genes. Humans and chimps share blood groups in the ABO system; humans and owl monkeys have very similar systems of sex-linked colour blindness. At every level of investigation it seems that the genetic links between ourselves and the other primates are remarkably close.

New staining methods have been used to compare the *chromosomes* of humans and primates in the hope of establishing patterns of relationship.

▷   Why are chromosomes likely to give only a crude insight into evolutionary relatedness?

▶   Each chromosome is made up of many thousands of genes so that their structure gives only a general picture of genetic change. For example, patients with whole sections of the α-globin gene family missing show no change in the appearance of chromosome 16, where these genes are located (Chapter 1).

Modern methods of staining make it possible to see more than a thousand distinct bands in the human karyotype. Every one of the bands in the human chromosome set is also present in chimps (Figure 5.3). Even orang-utans, which are less closely related, have a pattern of chromosome bands similar to our own.

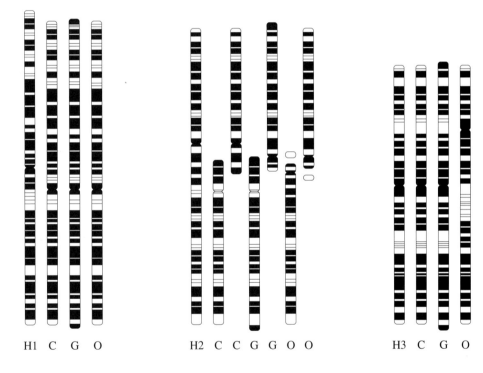

H1   C   G   O   H2   C   C   G   G   O   O   H3   C   G   O

Figure 5.3   Structural similarities and differences of chromosomes 1–3 of humans (H), chimps (C), gorillas (G) and orang-utans (O).

There seems to be almost no difference in the amount of genetic material needed to make a chimp compared to a human. There has, though, been some rearrangement of the order of chromosome bands in the two lineages since they separated, through the accumulation of inversions and translocations (Chapter 4). Humans are unique among primates in that two ape chromosomes are fused together to form human chromosome 2. There are also a few sections of highly repeated DNA (which forms characteristic '**C-bands**' in humans) which are not present in chimps and gorillas. These three species are so similar that it is difficult to make a convincing evolutionary tree showing their relatedness from chromosomes alone. One showing the possible connections between the old world monkeys and apes is shown in Figure 5.4. It is notable that some lineages have undergone much more chromosome reshuffling than others; why, is not known.

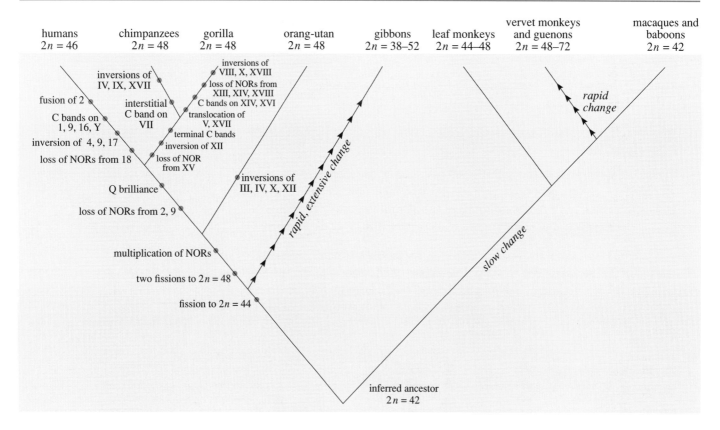

Figure 5.4 An evolutionary tree of the primates based on chromosomes. The tree shows the major chromosomal alterations that have occurred. Each blue dot represents an event inferred from the karyotypes of living species. Within any branch the sequence of dots is arbitrary. NOR = nucleolar organizer (cluster of actively transcribed rRNA genes within the nucleolus), Q brilliance = changes in heterochromatic (see Chapter 2) portions identified by staining with quinacrine.

Results from structural comparisons of *proteins* also hint at the closeness of the relationship between humans and other primates. Many have been sequenced and, again, the general picture is one of similarity. For the α- and β-globin chains considered together, for example, chimps and humans are identical while gorillas differ at just two of the 287 amino acids in these two chains (141 in the α chain plus 146 in the β chain). Table 5.1 shows how much the sequence of β-globin is conserved in other primates. Figure 5.5 shows evolutionary trees of relatedness constructed from patterns of amino acid sequence.

Table 5.1 The variable sites in the β-globin chains among some primates. All the other sites in the chain are shared.

Primate species	Amino acid number			
	80	87	104	125
human	Asp	Thr	Arg	Pro
chimpanzee	Asp	Thr	Arg	Pro
gorilla	Asp	Thr	Lys	Pro
orang-utan	Asp	Lys	Arg	Glu
gibbon	Asp	Lys	Arg	Glu
old world monkeys	Asp	Glu	Lys	Glu
lemurs	Glu	Glu	Thr	Ala

To make a complete (or even a partial) sequence of human and primate DNA – which is what would be needed to make an comprehensive statement of how closely they are related – will take many years. In the interim, DNA hybridization (Chapter 2), a crude but effective way of comparing the overall similarity of DNA molecules has given the strongest hint of how similar we are to the rest of the living world.

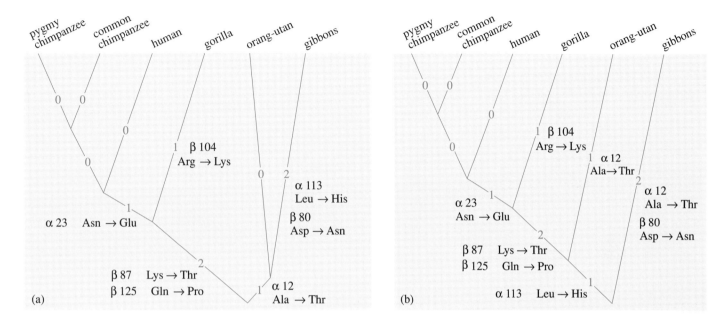

Figure 5.5 Examples of primate evolutionary trees based on amino acid substitutions at numbered positions along the α and β chains of haemoglobin. (a) Lowest number of amino acid replacements needed. (b) Amino acid replacements suggested by the overall molecular evidence available.

The method (Box 5.1) depends on the extraordinary toughness of the DNA molecule and on the desire of single DNA strands for togetherness – to bind to a complementary strand bearing a sequence that matches their own. When double-stranded DNA in solution is boiled, the bonds between the A–T and G–C pairs 'melt' and two single strands appear. As the liquid cools, these solitary strands collide with each other and – whenever complementary sequences meet – they reassociate into the double-stranded form.

▷ Why should the extent of pairing of DNA single strands from different species reflect the evolutionary distance between them?

▶ Because the stability of a DNA duplex (double helix) depends on the precision of base-pairing. The more similar the sequence in each – that is, the more closely related they are – the more stable the hybrid molecule.

As explained in more detail in Box 5.1, the stability of the hybrid DNA determines how well it copes with high temperature: duplexes with weak pairing due to poor sequence matching melt more easily. Sequence similarity can hence be measured simply by taking the temperature at which half the single strands have reassociated. Comparing that measure in 'hybrid' versus 'single-species' experiments gives a hint as to how much change there has been since the species involved split apart.

Box 5.1 *The DNA hybridization technique*

DNA labelled with radioactive iodine is mixed with non-radioactive DNA from the same species or from a different species. The mixture of double-stranded DNA is boiled for 5 minutes to separate it into single strands. On cooling the mixture to 60 °C for 120 hours four types of DNA are produced: (1) double-stranded and non-radioactive (that is, the same as in the original sample, reconstructed); (2) single-stranded; (3) double-stranded, with both strands radioactive (a very small proportion); (4) double-stranded hybrids – only one strand radioactive (Figure 5.6). The mixture is then placed on a column of hydroxyapatite. This substance binds double-stranded DNA but allows single strands to be washed through. The temperature of the column is increased step by step and the radioactivity present in successive samples of the solution emerging from the column measured. The radioactivity in each of the samples shows how much the hybrid duplexes have melted at each temperature. These values are plotted as **melting curves** of dissociation against temperature. Different curves will be produced for homoduplexes (DNAs both from same species) and heteroduplexes (DNAs from different species). The difference is often calculated as a comparison of the temperatures at which dissociation into single strands is half complete. This is referred to as the T_{50H} (the temperature at 50% hybridization). The difference in this parameter, ΔT_{50H}, between two species is an indicator of how much their sequences have diverged since they shared a common ancestor.

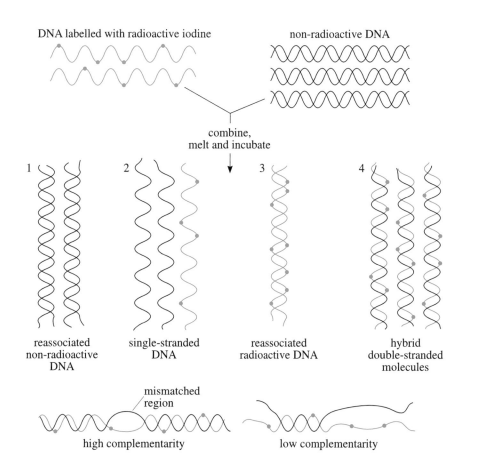

Figure 5.6 Formation of DNA hybrids – DNA strands from closely related species will have high complementarity (shorter mismatched regions) than those from more distantly related species and this is reflected in the corresponding melting curves.

The results for primates are remarkable. Chimps and humans share 98% of their DNA sequence. Gorillas are slightly less closely related. Orang-utans, gibbons and the old world monkeys show increasing levels of divergence (Figure 5.7). Although the differences between humans, chimps and gorillas are too slight to be sure about the details of the relationship it is clear that, in evolutionary terms, we are close relatives.

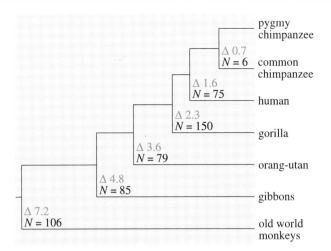

Figure 5.7 Patterns of relationship among primates based on information from DNA–DNA hybridization.
Δ = average ΔT_{50H} values (see Box 5.1); N = number of DNA–DNA hybrids that were averaged.

5.5   Dismantling the Darwinian machine

5.5.1   Genetic variation in nature

Given that there is a struggle for existence (which is true for most organisms) the mechanism proposed by Darwin has three parts: mutation (which gives rise to diversity), natural selection and random change. Ultimately, these processes may lead to the interruption of genetic exchange between two groups – to the origin of species. Each is illuminated by the new genetics.

All variation comes, ultimately, from mutation; originating either in the parents of the existing generation or in ancestors long dead. The process itself is described in Chapter 4. Here we discuss its longer-term effects; the accumulation of genetic diversity or polymorphism (Chapter 1) within populations.

It was once assumed that most new mutations were harmful. Natural populations must, it seemed, contain little genetic variation; most mutations disappeared at once, and those few favoured by selection soon replaced their ancestors. At one time, this appeared to be a problem for Darwinian theory: if there was no variation, how could evolution happen so readily? With improvements in technology, more and more polymorphism has emerged. Now, if anything, the problem is the opposite; there appears to be too much diversity to allow it all to be influenced by selection.

At first sight, most creatures are not very variable. The whole of biological classification – taxonomy – depends on shared similarity within species and differences among them. The few instances in which there is a large amount of visible variation became classics of evolutionary genetics. Plate 5.1 shows some variants in shell characters in the polymorphic land snail *Cepaea nemoralis*. There are inherited differences in shell colour and in the number and appearance of shell bands, giving about 20 000 possible phenotypes. There are similar instances from a few butterflies, some marine organisms and, to a lesser extent, from humans.

These are exceptions. Physical uniformity is the rule. Most fruit-flies – and most chimpanzees – look much the same, at least to human observers! Where, then, is the raw material upon which selection can work?

In fact, the problem is not a real one. Under this phenotypic identity is a mass of diversity. It can be uncovered in many ways. Figure 5.8a shows the pattern of wing veins on a normal *Drosophila* wing. Occasionally individuals are found in which the cross-veins are interrupted (Figure 5.8b). The condition – *crossveinless (cve)* – is heritable. Because it is rare, *cve* might seem to be one of those mutations that turn up now and again and are quickly removed by selection.

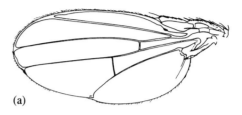

(a)

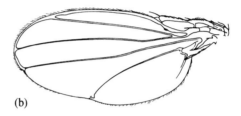

(b)

Figure 5.8 (a) Normal *Drosophila* wing. (b) Wing of *crossveinless (cvl)* fly

A simple experiment shows that this is not true. Most flies carry genes predisposing them to making wings without cross-veins. All that is needed is to heat the flies up briefly as pupae. Up to half the flies may then produce *crossveinless* wings. The heat shock opens the window to a hidden world of polymorphism.

▷ What kinds of environmental shocks might help uncover hidden genetic diversity in humans?

▶ New chemical challenges – drugs, foods and medicines – have exposed many. Sensitivity to suxamethonium due to variation in the enzyme acetylcholinesterase is but one example (Chapter 4). There are also individual genetic differences in the ability to metabolize recreational drugs such as alcohol.

Molecular biology has provided the tools that make it possible to look directly at polymorphism in the genome rather than inferring its existence by exposing organisms to environmental stresses. Every technical advance has uncovered a new range of variation.

The blood groups were first: mixing blood from different people showed that every individual had his or her personal set of antigens. Now, antigenic diversity on the surface of cells can be seen to be so pervasive that everyone (except for twins) is unique. Gel electrophoresis of proteins (Chapter 4) came next: looking at inherited variation in a sample of gene products using changes in molecular size and charge which result from the substitution of one amino acid by another. The first test of human protein-coding loci showed that about a quarter of the genes were polymorphic.

▷ This figure is now known to be an underestimate of the real amount of variation in this sample of genes. Why should this be?

▶ Because of limitations in the technique of electrophoresis. The method picks up differences in the shape and charge of protein molecules which alter the extent to which they migrate into the gel. Many individual differences involve amino acids with the same charge and are missed.

This estimate was so much in excess of previous guesses at the amount of variation in nature that many geneticists disputed that a quarter of all genes could possibly be polymorphic. The proteins used were all enzymes: understandably so, as enzymatic activity is what was used to identify their presence on the gel. Some other proteins are, indeed, much less variable. There is almost no diversity in, for example, the crystallin protein of the eye or the ribonucleoprotein which makes up much of the ribosome.

▷ Why are there such differences in the extent of polymorphism among different classes of proteins?

▶ Any protein that has a structural role and that must link up with other subunits to do its job is likely to be less tolerant of new mutations than are those with less exactly defined functions. Crystallin molecules, for example, are arranged in a precise array within the eye: any variant would not fit into the lattice and its carriers do not survive to pass it to the next generation.

The study of polymorphism in proteins has been superseded by DNA technology. Using a variety of restriction enzymes (Chapter 2) DNA can be obtained from the genetic material of two unrelated people. Comparison of the DNA shows that about one DNA base in a thousand differs between them. This gives, at a very crude estimate, around three million potentially variable sites in the human genome. Most of these are in non-coding segments, and – as we saw in Chapter 2 – many arise from individual differences in the position and number of repeated lengths of sequence scattered throughout the DNA. Whatever the details, it is clear that Darwin's claim that natural populations are full of inherited variation has been vindicated by molecular biology.

5.5.2 Natural selection as the engine of evolutionary change

Darwin saw *The Origin of Species* as merely an outline for a much fuller work on natural selection, the core of his theory. Ironically enough, the new genetics has both provided elegant examples of selection in action and cast doubt on whether this process is generally important in controlling the structure of genomes or the genetic differences among species.

The first chapter of *The Origin* – and an important part of Darwin's 'one long argument', as he described the book – turns on the successes of farmers in producing new varieties of plants and animals by artificial selection. The case made there that choosing the best and discarding the worst can lead to rapid change – to evolution – is so strong that the reader finds it easy to accept the more speculative examples that Darwin later gives from nature. Exactly the same is true for selection at the molecular level: some of the most convincing examples come from artificial evolution in the laboratory.

RNA is a remarkable molecule. It functions both as an enzyme, catalysing reactions in the chain from DNA to protein and as a replicator, containing the information, coded into nucleotides, to make copies of itself. In RNA viruses, of course, it carries out both these functions. For this reason, RNA may have been the molecule present during the early evolution of life itself: DNA is a replicator but not an enzyme, and proteins are enzymes but not replicators. An early **ribozyme** (Book 1, Chapter 5) able to catalyse its own propagation, would have spread rapidly in a world where the chemicals for life, but not life itself, were present. Natural selection would favour anything that increased the rate and accuracy of proliferation. DNA probably came later.

By applying artificial selection to RNA it is possible to mimic part of this process. The protozoan *Tetrahymena* has an RNA that is good at cutting and splicing other RNA molecules. Under very artificial conditions (high temperature, with lots of magnesium ions present) it can be persuaded to cut DNA as well, albeit inefficiently. Under normal circumstances, it is almost unable to do this. Into a population of many millions of identical RNA molecules, mutations were introduced by chemical means or by multiplying the RNA molecules using a modified polymerase chain reaction (Chapter 2) to make imperfect copies. Those few mutated RNAs that were able to cut the DNA even slightly were picked out and amplified in number, using the ability of RNA to generate a complementary cDNA. This can itself can make thousands of new copies of its template RNA. The process was repeated in ten successive 'generations' of ribozyme molecules.

This process of mutation and artificial selection led to the rapid evolution of a new kind of RNA; one that could cut DNA up to 100 times more efficiently than its ancestor, ten generations earlier. The experiment was carried out several times, and there was some consistency among the replicates in the mutations incorporated to make the new DNA 'super-catalyst'. Figure 5.9 shows the course of evolution under artificial selection over the course of the experiment. The new RNA, although it has never (as far as one knows) been seen in

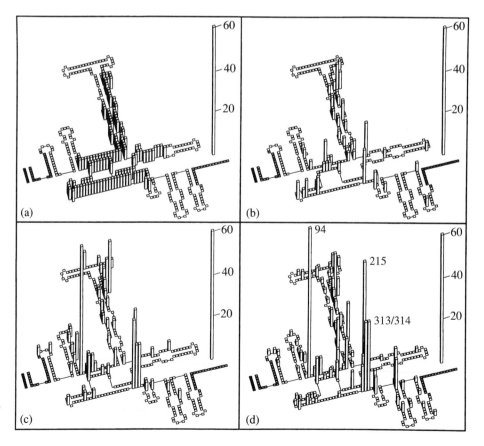

Figure 5.9 Evolution of a new catalytic property of RNA using artificial selection. The sites at which mutations occurred over the course of evolution are superimposed on the secondary structure of the *Tetrahymena* ribozyme. Box height corresponds to mutation frequency (%) – see scale – at each nucleotide position. Non-mutatable primer binding sites are shaded grey; RNA substrate is shown in black. (a) generation 0; (b) generation 3; (c) generation 6; (d) generation 9. The labelled positions in (d) are sites of frequent mutation.

nature, has real potential: bacteria into which it is inserted are able to resist infection by DNA bacteriophages.

This experiment leaves a diverse group of molecules facing a new environmental challenge. Only those relatively more able to meet it – in this case to cleave DNA – survive and copy themselves. Exactly the same process goes on in nature. Mutations which increase the efficiency of the enzyme acetylcholinesterase have spread world-wide in insects since the development of modern insecticides. The diversity of evolved responses to malaria infection seen in the structure of haemoglobin (Chapter 1) is testimony to the efficiency of natural selection in leading to rapid change since malaria first began to affect humans, probably only about 10 000 years – 500 generations – ago.

5.5.3   Evolution by accident

Darwin was well aware of the importance of random change in evolution. The unique land snails on Madeira and the finches on the Galapagos Islands helped formulate his idea that creatures sharing common descent could diverge. It is no accident that islands are fertile fields for genetic change. As well as their unique range of habitats – and the unique opportunities for the first invaders to exploit them – oceanic islands are isolated. Only a few animals and plants ever get there. Frogs, with their need for fresh water, almost never do: and this was part of Darwin's case that local evolution – rather than special creation – explains the diversity of life.

If only a few founders make it to an island they will inevitably carry only a sample of the variation present in their ancestral population. A single female fertilized by a single male, for example, carries only four of the thousands of genomes present in her home population. Her offspring can only possess these four. By going through a bottleneck in this way, the founder group undergoes genetic change – it evolves. This evolution happens purely at random, by the statistical accidents of sampling. It has nothing to to do with natural selection. Darwin accepted that this could happen, but had no way of knowing how important such events might be.

We now know that genetic bottlenecks are a major force in evolution. The fate of any allele depends on the size of the population in which it finds itself. In a small population, random change may lead to a sudden increase or decrease in the abundance of a particular allele as individual carriers succeed (or fail) in reproducing; perhaps, even quickly replacing what went before. In a larger population alleles take longer to change frequency. A knowledge of population size is crucial in judging the importance of such 'genetic drift.' This means not just its size today: a reduction in numbers long ago (such as the arrival of a few founders in a new place, or a catastrophic environmental change) can have persistent effects.

For most creatures, this poses a problem as we know nothing of their demographic history – whether they have indeed been through genetic bottlenecks in the distant past. Humans are unique in that they keep records and leave relics of their history. It is easier to study the effects of changes in population size in our own species than in any other.

The classic instances of the role of random change in animal evolution come from Pacific islands – the Galapagos and (an even better example of the importance of isolation) the Hawaiian archipelago, which has enormous numbers of endemic, that is, locally unique, species of insects, birds and plants. Appropriately enough, one of the best examples of the importance of genetic bottlenecks in humans follows the same pattern – the spread across the Pacific and the progressive loss of diversity as one population bottleneck succeeded another.

Human populations contain many lineages of mitochondrial DNA (described at greater length in Chapter 6). This is part of the genome, but is carried in the cytoplasm. As a result, it is inherited *matrilineally*, through the egg. This means that mitochondrial DNA contains the history of the world's women. Because it has no repair enzymes, mitochondrial DNA quickly accumulates mutations. It has been much used, therefore, for studying evolutionary history.

There are two branches in the mitochondrial lineages of the peoples of the Pacific. One group has a deletion of a nine base-pair sequence in the mitochondrial DNA molecule. Mitochondria with this genotype are found in East Asia and in much of Oceania – the distant islands of the Pacific, to Hawaii and New Zealand. The other lacks the deletion, and fills Australia and Highland New Guinea. This fits the known history of human colonization: Australia and New Guinea were populated 50 000 years ago (before Europe, incidentally) by invaders travelling down the chain of large islands which link them to the Asian land mass. The modern peoples of Oceania, though, are descendants of a later wave of farmers that originated in East Asia with the population explosion that took place when rice farming started 10 000 years ago. They spread slowly across the ocean, reaching its most distant islands, Hawaii and New Zealand, only about 1 000 years before the present.

The number of mitochondrial lineages is less in the remote islands of the Pacific than in Papua New Guinea or even in modern East Asia. This is because of the small numbers of pioneers (in particular the small number of women) who founded each island, and the restricted numbers of people who lived on the islands in the generations after it was first colonized. Sometimes, the effect is dramatic. All Maoris, for example, share a common mitochondrial genotype: all can trace a descent from a common female ancestor.

This does not, of course, mean that all modern New Zealanders descend from the only woman to step ashore from the founding vessel a millennium ago. There may have been tens, scores or even hundreds of females who made it to that empty land. However, should any one of them (or her female descendants) have failed to have a daughter, then her mitochondrial lineage will be extinct. A long period of moderately small population numbers, with the accompanying loss of a few such lineages each generation, can have the same genetic effect as does a dramatic episode involving just one or two founding mothers.

Genetic drift, either through sudden bottlenecks or persistent small population size, has certainly affected human evolution. In the recent past, founder effects have had persistent effects on the distribution of genetic disease. Many of the patches of Huntington's disease across the world can be traced to a single individual in historic times. All those carrying the allele in the area around Lake Maracaibo in Venezuela (where hundreds of people are affected) are descendants of a woman called Maria Concepcion, who migrated there more than a century ago (see Video sequence 4). Huntington's disease – and other genetic ailments such as porphyria, an inherited failure to metabolize the breakdown products of haemoglobin – is also relatively common among Afrikaners: it is no coincidence that the millions of Afrikaners alive today nearly all descend from a hundred or so families (some of which must have carried genetic disease) who moved there during the 17th and 18th centuries.

The effects of genetic drift may echo through history for much longer periods. In general, Africans south of the Sahara (which is where modern humans originated) have higher levels of heterozygosity at the DNA level than do most of the other peoples of the world. Perhaps the reduction in diversity outside our natal continent reflects ancient population bottlenecks as humans first entered the Middle East and began to spread across the globe.

The importance of genetic drift in evolution remains a matter of contention. It certainly has some effect on all new mutations (apart from lethals), however advantageous or disadvantageous they might be. There must have been many cases where an advantageous mutation appeared in someone who, purely by accident, failed to pass it on. Sickle-cell disease, for example, has just a few foci of origin (Chapter 1); but, given the known mutation rate and the numbers of people who have lived in malarious areas over the past 10 000 years, the new allele must have appeared by mutation on thousands of occasions. Most of those who inherited it, though, failed to pass it on, either because they died for other reasons or through the probabilistic nature of Mendelian segregation. Just a few of the new sickle-cell genes made it through the initial bottleneck and were picked up and spread by natural selection.

Random change is most likely to affect genes whose fate is not much moulded by selection. It probably explains much about evolution at the level of the DNA sequence. For many base substitutions – those at the third position of most codons, or within pseudogenes – it is hard to see how selection could be effective as no gene product is involved. Such mutations may appear and disappear at random. Some believe that many other gene substitutions – including the majority of protein polymorphisms – also arise and disappear through the accidents of sampling. Others dispute this **neutral theory** and argue that natural selection acts at effectively all gene loci. The issue, as is true for many others in evolutionary genetics, remains unresolved.

5.6   Genes as clocks – dating evolution without fossils

The family tree hidden in the genes can do more than confirm the fact of evolution. If there is, within a particular group, a landmark which can time when two of its evolving lineages separated (perhaps a fossil, or a dated geological change which erected a barrier between them) it may be possible to test the rate at which other members of the group have changed, and the date when they split, by comparing their genes with those whose history has been established from geological evidence. This is the basis of the **molecular clock**.

The AIDS virus has the unique advantage that its clock can be calibrated with dates from the history of the infection's spread; for the Florida dentist's patients, for example, the day when they had their teeth filled. In general, the time of infection fits well with the idea that the accumulation of mutations in the virus does indeed act as a chronometer, ticking at a more or less steady rate. Comparing the virus in two people exposed to a common source can hence give testimony as to when they were each infected even when no other information is available.

The molecular clock has had some spectacular successes in timing the past. Perhaps its greatest triumph has been in the primates. Before the advent of molecular biology the evolutionary gap between chimps and humans was thought to be huge. The primate fossil record is so poor that some palaeontologists felt that it was possible that chimps, humans and gorillas last shared a common ancestor as much as 20 million years ago.

The molecular data changed views of the age of our own species. It is difficult to calibrate the primate clock because of the incompleteness of the fossil record of our early ancestors. There are, however, a few well-dated fragments from the primate past. A fossil from upper Egypt which might be close to the common ancestor of all old world monkeys and apes has been dated to around 33 million years ago. The first ape lineage appears around 10 million years later in the Rift Valley of East Africa, and there is a hint of an early orang-utan in the same region around 15 million years before the present.

Plugging in the data from proteins and DNA from monkeys, apes and the orang-utan to these fragmentary pieces of geological information gives a date for the last common ancestor of humans, chimps and gorillas of only about six million years ago. As our biological split from other apes was so recent, it is very clear that whatever makes us human does not reside only in the DNA. Early in our history as a separate species, cultural evolution emerged as an independent force for change: the machine that can measure the evolution of culture has not yet been invented.

Irrespective of the strengths of the clock, it does have its weaknesses. Biologists' confidence has been shaken by the realization that the rate at which the molecular clock ticks can speed up or slow down; and that even when its hands move more or less regularly their speed can vary in different lineages. Just as in physics – where it is often necessary to imagine improbable things happening to clocks on spaceships or at the beginning of the Universe – the optimistic early days of molecular timekeeping (when dating by counting genetic change was credited with almost miraculous accuracy) have gone. The clock hidden in the genes has begun to look more and more relativistic.

First, its rate varies with the molecule used. Some proteins – those of the ribosome, for example – scarcely change over millions of years, while certain DNA sequences (such as those within introns or pseudogenes) evolve far more rapidly. Part of this difference is due to the 'policing' role of natural selection, removing new mutations from crucial molecules as they arise, while ignoring the capricious behaviour of parts of the genome with few apparent functions.

The Isthmus of Panama rose above the sea about three million years ago. It allowed North American mammals to invade the South and split the marine life of the Pacific from that of the Caribbean. Since then, many sea creatures have diverged into new – but closely related – species on either side of the bridge; an ideal chance to test the idea of a molecular clock and of how well it works in different parts of the genome.

Four closely related pairs of sea urchin species were tested. One clock came from variation in restriction fragment patterns in mitochondrial DNA. Variation at loci coding for variable proteins was also examined. This is a different sample of the genome, representing only coding gene loci and mutations which produce changes in amino acid sequence (Figure 5.10).

Patterns of mitochondrial DNA divergence were quite consistent. There was at least ten times as much difference between the members of each pair as within them, and the rate of divergence of mitochondrial DNA was almost the same in each species, at about 2% per million years. However, the proteins were much less regular, as one pair of species had scarcely changed at all (Figure 5.10, top). Why this is so is not certain; perhaps natural selection removed new variants as they arose. Whatever the reason, it is clear that there is no universal clock in these creatures. The rate at which it ticks depends on the molecule used.

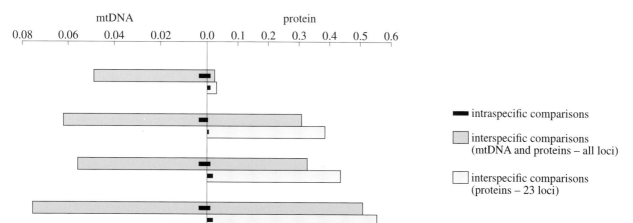

intraspecific comparisons

interspecific comparisons
(mtDNA and proteins – all loci)

interspecific comparisons
(proteins – 23 loci)

Figure 5.10 Mitochondrial DNA (mtDNA) and protein clocks for creatures separated by the Isthmus of Panama. The figure shows estimates of divergence within and between four related pairs of sea urchin species. Each set of bands represents estimates for mtDNA and proteins for one species pair. The numbers on the scale represent units of divergence.

When the same molecule is compared in different lineages, too, molecular clocks go more rapidly in some groups than others and can speed up or slow down. This is, of course, well known to be true for morphological evolution. Some groups change rapidly, others not at all. Frogs are, basically, just frogs: they may differ slightly in size, in song or in how sinister a poison they produce, but beyond that they are (except to herpetologists) a singularly monotonous group. Mammals, on the other hand, include whales and dormice, bats and moles, humans and pandas. If we had no fossils we might guess that the mammals must have been around for far longer than the amphibia. How else could they have produced this huge diversity? Surely they must have had far longer in which to do it.

The opposite is, of course, true. Fossils show that frogs have lived on Earth for at least ten times as long as mammals. Perhaps there is something about being a frog which makes it difficult to evolve – or perhaps frogs are so well adapted to the narrow range of habitats in which they live that they do not have much need to reinvent themselves.

Whatever the reason behind the conservative frogs or the radical mammals, any attempt to date the evolution of the two groups by comparing morphological differences among living species would be misleading. There are analogous differences in the rate of evolution at the molecular level, too. This means that, just like changes in body form, clocks should not be accepted uncritically as a statement for how long evolution has proceeded.

When comparing the rate of evolution in two related groups, it is best to use an 'outgroup' – another, distantly related, reference lineage – against which both can be compared. When the rate of change in various vertebrates is measured in this way, marked differences appear. Sharks have a slow molecular clock. Their fossil record is good and, as a result, we can be confident about the major events in shark evolution. Comparison of the amino acid sequences of sharks which have been separated for tens of millions of years show that they have changed far less than have, for example, chimps and humans, who shared a common ancestor six million years ago.

Even on shorter scales, the clock speeds up and slows down. The rate of molecular change in the lineage leading to humans and the great apes from their shared ancestor with other old world monkeys is slower than that of its other branch (which leads to the monkeys of the old world). In fact, the clock ticks twice as fast in other primates compared to the human/great ape family. Among rodents, its hands move even more quickly. Why should this be?

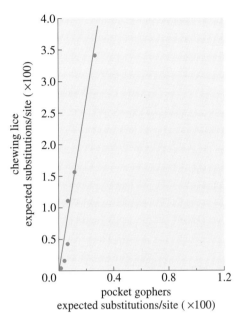

Figure 5.11 Comparison of the rate of molecular change in a region of a gene from chewing lice and their pocket gopher hosts. Notice that there is an approximately 11-fold difference in rates of accumulation of synonymous (silent) DNA substitutions in the two species.

Nobody really knows. Some of the differences arise because evolution works in generations rather than in years: mice reproduce rather more often than humans do. Their DNA has been through more rounds of replication (with their opportunities for making mistakes) during the formation of sperm and egg than has the genetic material of more phlegmatic creatures such as ourselves – or sharks. A neat example of the importance of generation times comes from comparing the rate of molecular evolution in another group of rodents, the pocket gophers, with the chewing lice that use them as a host. Every gopher species has its own species of louse, so that evolution in the two groups has taken place – as it often does in host–parasite systems – more or less in parallel. DNA sequence change, though, is much faster in the lice (which have many generations to every one in the host). The rate of accumulation of synonymous DNA substitutions – those causing no change in amino acid sequence – is more than ten times higher when related louse species are compared than in their equivalently related hosts (Figure 5.11) This reflects an approximately ten-fold difference (with a six-week versus a one-year life cycle) in the number of generations since the common speciation event in the parasites and their hosts.

Mice and humans differ in other ways, too. Mice are smaller, for one thing. Among vertebrates as a whole, there is a tendency for large creatures to evolve more slowly at the molecular level than do small. Part of this is due to the generation time effect; but part may be due to differences in metabolic rate. Creatures that live relatively fast have comparatively swift molecular clocks. Sharks – with their slow and cold-blooded way of life – and the fast-living primates have roughly the same range of generation times, but shark mitochondrial DNA changes at only a fifth the rate of that of primates. Perhaps the waste products produced by animals with a more agitated life-style damage their DNA and speed up their molecular clock.

There are other possibilities, too. Perhaps the slowing of the clock in the lineage leading to humans reflects how good we are at altering our environment to suit our needs – but this is a guess.

The hope that there would, one day, appear a universal molecular clock which could be used to place every creature in its evolutionary context by sequencing a segment of its DNA was over-optimistic. Instead, there seems to be a range of local clocks, each appropriate to the particular gene and taxonomic group being studied. Darwin himself was hopelessly wrong in estimating the speed of evolution from the geological record and from comparative anatomy: the Earth was far older than he thought. Molecular biology has shown that we are all living fossils, containing within ourselves a record of our past. The finding that the evolutionary paths of humans, chimps and gorillas split only about six million years ago would never have been accepted if it were based on a few fragmentary fossils. An imperfect clock is better than no clock at all.

5.7 Evolution and the size of the genome

Evolutionists are used to seeing beautiful and ingenious solutions to the problems that life presents: the albatross, the blue whale and even the tapeworm are efficient and economical adaptations to the places where they live. It once seemed reasonable to hope that the genome would show the same fit between form and function. However, it did not take long to discover that the structure of DNA is complex, surprising and – still – fairly inexplicable. Some of its attributes fit perfectly into the Darwinian framework. With a certain amount of ingenuity others can be made to conform; but genomes also have properties that demand modes of change quite alien to those working on the biology of whole organisms.

The size of animals makes sense to a Darwinist. Blue whales are bigger than amoebae for obvious adaptive reasons. The size of their genomes is harder to understand: an amoeba genome is a hundred times bigger than that of a whale (Figure 5.12). Mammals, from mice to whales, differ only by about four times in the number of DNA bases they contain. Bony fish, in contrast, vary by more than 300 times in DNA content from species to species. Even rather similar creatures vary greatly in genome size: *Drosophila melanogaster* has about 200 million bases, but the locust has 50 times as many. How can this *C-value* paradox (Chapter 2) be a response to natural selection?

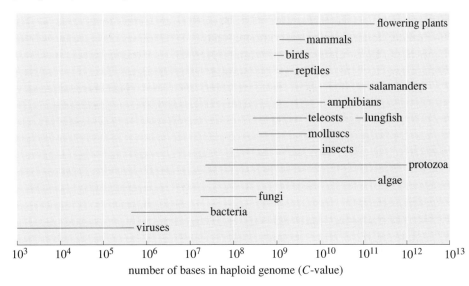

number of bases in haploid genome (*C*-value)

Figure 5.12 The haploid DNA content (*C-value*) in various groups of organisms.

To make things worse, there is no clear fit between the amount of DNA and the number of working genes. Estimates vary from about 3 000 protein coding loci in yeast to around 100 000 in mammals – a range of less than 50-fold, compared to the 80 000-fold difference in the amount of DNA per genome.

Most of the profusion of DNA in locusts or in amoebae does not contain any genetic information at all. In many cases, it is made up of hundreds of thousands of copies of the same sequence. Just what it does is one of the central questions of genetics. Perhaps it is just the debris of the past – the remnants of ancient invasions by parasitic DNA, for example – but perhaps it has some crucial function of which we know nothing.

As described in Chapter 2, much of the genome has a duplicated structure. Often, the duplications are palindromes, in which a short sequence is partnered by another with the bases arranged as a mirror image. Highly repetitive DNA consists of short sequences, up to a few hundred bases long, repeated hundreds of thousands of times. Moderately repetitive DNA has longer segments, perhaps a few thousand bases long, repeated hundreds of times.

There are also considerable differences in the way that repetitive DNA is arranged within the genome. Some localized sequences sit next to each other in tandem array. Occasionally, as in the kangaroo rat, there are huge numbers of adjacent copies – 2.4 billion repeats of the short triplet AAG, in this case. When extracted and separated according to weight in a density gradient centrifuge this mass of homogeneous DNA forms a thick band called **satellite DNA** (Figure 5.13). Often, as in some species of *Drosophila*, localized repeats colonize just one part of the genome, leaving the rest unscathed.

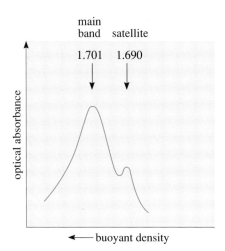

Figure 5.13 Separation of DNA into a main band and a *satellite* band by centrifugation through a density gradient of caesium chloride.

The interspersed sequences are scattered throughout the genome and, as you learnt in Chapter 2, exist in two main forms: SINEs (short interspersed repetitive elements) are usually less than 500 base pairs long and LINEs (long interspersed repetitive elements), longer than 5 000 base pairs. The number of each varies greatly, and there are similarities in sequence between LINEs from very different species – humans and *Drosophila*, for example.

New computer techniques for searching for patterns in the genome give intriguing hints of repetition on a larger scale. Some stretches of DNA, encompassing hundreds of thousands of base pairs, are rich in the two nucleotides G and C, followed by equally long sections in which A and T predominate. Sometimes, as in mammals, the patterning is strong enough to show up as distinct bands when chromosomes are stained (Chapter 2).

How did all this repetition, and the huge range of genome sizes, arise? Various processes of duplication – some ancient, some recent – must be involved. The whole of evolution is marked by themes and variations: an ancestral structure of some kind which has been multiplied and modified. This can be seen in body structure (think of the earthworm) and in gene families such as the haemoglobins (Chapter 1). Studies of the structure of the genome as a whole suggest that duplication is everywhere.

▷ Taking the simplistic view that ancient genomes were as small as those of modern bacteria (around 4 million base pairs), and that the mammalian genome (3 000 million base pairs) is the most evolved of all, how many rounds of duplication were needed to expand the ancient genome into the modern one?

▶ Between ten and eleven rounds of duplication are enough to get from bacteria to mammals.

These very ancient duplications may help to explain the long-range repeats seen in most large genomes. There have, however, been other multiplication events of various kinds. They are very different in nature, and their relative importance in genome evolution is scarcely understood. However, all have at least the potential to increase genome size.

Polyploidization – the multiplication of complete genomes – is common in the history of many species (Chapter 4). Chromosomes show that grasses and wheats and fish such as salmon and trout have duplicated their genomes relatively recently, either by doubling up their own chromosomes or by absorbing a complete set from a closely related species.

There are hints of such doublings in the mammalian genome too. Long sections of working genes – linkage groups – are remarkably similar in sequence on different chromosomes. For example, human chromosome 12 carries a linked group of genes coding for keratin (the hair protein), for collagen (a major constituent of skin) and a homeobox (a control region important early in development – Chapter 3). Chromosome 17 has exactly the same group of genes, changed only slightly in function from their equivalents on an apparently independent chromosome. This suggests that there was, early in the history of vertebrates, a round or two of multiplication of the complete chromosome set. Such processes probably help to explain part of the general increase in size of genomes during the history of life.

▷ What types of gene might be expected to be particularly liable to duplication?

▶ Those for which there is a high demand for their specific gene product at some time during development. For example, there may be fifty copies of the gene for insect eggshell protein and several hundred for the nuclear genes responsible for producing rRNA. Sometimes, duplication can take place rapidly. Resistance to organophosphate insecticides by aphids and other insects may involve the production of many repeated copies of a particular esterase gene within only a few generations.

Duplication is often followed by divergence as the existence of a second copy of particular gene provides the raw material for a new evolutionary departure.

Most animals have, early in development, a segmented body plan. Sometimes, as in the earthworm, this remains obvious. More often, as in mammals, a basic simplicity of design is confused as each element develops into a specialized adult structure. These themes – repetition and divergence of structure – also exist in the genome. Gene duplication may be the main way in which new adaptive functions evolve.

Many genes are arranged into **gene families**, DNA segments that are similar in sequence but have diverged in what they do. Sometimes (as within each of the two globin gene families – Chapter 1) the members are adjacent in the genome; sometimes – as in the α and β families themselves – they have been split apart during evolution and are separated (or even scattered all over the chromosomes). Often, their DNA sequences are so alike that common ancestry is obvious, but on occasion two genes in a family can be seen to be related only on the basis of subtle similarities in structure. Their duplication must have taken place long ago. Comparing patterns of sequence change within members of a particular gene family since their duplication gave the first hint of some surprising new evolutionary forces at work.

Human growth hormone and prolactin (the hormone responsible for milk secretion) are not, at first sight, very alike. Although both are produced in the pituitary, they differ in what they do, and in about a third of their 190 or so amino acids. At the DNA level they are more similar, as each is interrupted by four introns at almost exactly the same place along the molecule. Other genes active in quite different tissues – such as placental lactogen (another hormone involved in the control of milk production) – have a DNA sequence close to that of these two.

The correspondence in structure of prolactin and growth hormone suggests that they are relatives: that each diverged from a common ancestor long ago. As all vertebrates – including amphibians and fish – have both of them (although prolactin does a different job outside the mammals) the ancestral DNA sequences must have doubled up at least 400 million years ago. Since then they have strayed apart in the genome: in humans the prolactin gene is on chromosome 6 and growth hormone on chromosome 17.

Placental lactogen, the other relative in the family, is found only in mammals, so that a second duplication must have taken place at or near the origin of mammals, around 100 million years ago – long after the duplication that founded the gene family and split prolactin and growth hormone. The placental lactogen gene has stayed near its parent, the growth hormone gene on chromosome 17. In humans, there are two copies of the growth hormone gene and three of placental lactogen, all close together, suggesting yet another more recent round of duplication.

Although prolactin, growth hormone and placental lactogen are quite clearly related, descending from a common ancestor, the DNA sequence of each member of the group has diverged to a different extent. Given their probable origin at the beginning of the mammals and the rate of change in this gene family since the origin of the vertebrates several hundred million years earlier, human placental lactogen and growth hormone have diverged much less than the other evolving members of their gene family.

Fossils show that the vertebrates emerged about 400 million years ago and that the mammalian line appeared 300 million years later. Using the earlier date to start the growth hormone–prolactin clock gives a rate of about 1% sequence divergence between the two genes per million years. Placental lactogen appeared when the mammals, with their placentae, did. However, a clock started 100 million years ago, timing the divergence between human placental lactogen and human growth hormone, ticks at only half the rate of the older clock. Why should such closely related genes evolve at such different rates within the same lineage?

There seems to be, within an evolving line, something which causes its DNA sequences to conform to a common plan; that is, within the mammals, there has been a slowing of the divergence of placental lactogen and growth hormone and hence of the molecular clock which times it. The rate of change is well below that of the equivalent sequences held in a different lineage: the evolution of both human placental lactogen and human growth hormone has slowed down – they are evolving together.

Such 'concerted evolution' is a frequent theme in genetic change at the molecular level. There is a whole range of evolutionary processes which act only within particular species and cease or change direction once the genes involved are held in different species: extra modes of evolution which happen within species, not between.

This, needless to say, flies in the face of Darwinian theory. Darwin himself was a **uniformitarian**: most of the arguments in *The Origin* rest on the fact that mechanisms seen within living creatures are sufficient – given enough time – to explain large-scale patterns of evolution. Molecular biology makes it clear that Darwin was right, but only up to a point. To understand the structure of the genome we have to bring in a whole series of new evolutionary mechanisms.

5.8 The autonomous genome: genome evolution as an intrinsic process

The biggest revision of evolutionary theory demanded by molecular biology is the realization that the genome has properties which cause it to change – to evolve – without much reference to the world outside. The problem for Darwinians began with the discovery of so-called 'junk' DNA. It was hard enough to accept that much of a genome produced no useful product; but even more so to see that – in spite of the absence of an avenue through which selection might work to remove the mutations which must surely occur – large sections of this apparently redundant DNA retain a controlled and repetitive structure in which the same motif is repeated thousands of times without error.

The problems become worse when comparing DNA sequences in closely related species. The toad *Xenopus laevis* has a large amount of DNA – about 3 000 million base pairs per haploid genome; the same as a human being (and twice as much as a chicken). Much of this is made up of millions of repeats of the same short DNA sequence. With some imagination it might be possible to come up with an explanation based on natural selection: perhaps the repeated sequences are involved in stabilizing chromosome structure, or in coping with some unknown problem in a toad's life. But what is to be made of the fact that the closely related (and morphologically almost identical) species *Xenopus borealis* has an equivalently large genome, but based on a different repeat unit? Something has led to **sequence homogenization** within each species since they separated. Some unexpected mechanisms of molecular evolution must be involved.

▷   If these unforeseen evolutionary patterns in DNA emerge only *within* species, what does this suggest about the processes that produce them?

►   They are likely to be associated with *sexual reproduction* – recombination of various kinds – which, almost by definition, happens within and not between species.

DNA has, it seems, properties of its own which generate patterns that cannot be explained by classical theories of evolution. There are mechanisms intrinsic to sexual reproduction and DNA replication which reach their limits at the sexual barriers that mark the boundaries of species. Much evolution might hence arise from the inherent properties of DNA rather than from the interaction between the organisms transmitting it and the environment in which they live.

Molecular biology has provided many new mechanisms of this kind which must be accommodated into the evolutionary synthesis. Together they have produced the doctrine – still far from universally accepted – of the **fluid genome**: that, under a veneer of morphological stability, animals and plants undergo restructuring of large parts of their genetic material. This process of flux may mean that, in some creatures at least, segments of the genome continually turn over as one sequence or another sweeps through an evolving lineage to be replaced by the next. Most of the mechanisms involved entail some form of asymmetric recombination.

The symmetry of recombination is, of course, central to Mendelian genetics. Loci on separate chromosomes show independent assortment – each locus is autonomous as it passes down the generations. This independence is reduced for loci on the same chromosome; greatly so if the two are closely linked. Even for linked loci, though, there are cross-overs. In Mendel's experiments, these were always *reciprocal*: the recombinant chromosome contains the same amount of genetic material, albeit in a different arrangement, as its parents (Chapter 2).

Detailed studies of tightly linked loci suggests that things are not always so simple. Often, there is unequal crossing over; an asymmetric exchange between the chromosome strands, so that one gains material and the other loses (see Chapter 2). Sometimes, unequal crossing over can be seen at the chromosomal level – as is the case for the *Bar* mutation in *Drosophila* (Chapter 4). The same thing goes on at the molecular level, although more copies of a particular sequence can be generated than the mere three found in ultra-*Bar*.

There are other methods for producing long sections of repeated DNA. **Gene conversion** was discovered in experiments on *Neurospora*, the bread mould. Some of its inherited variants change the appearance of the sexual spores, the *ascospores*. Conveniently enough, when these are produced by meiosis each set is held for a while in an *ascus* (that is, in the same enlarged cell). The products of meiosis can thus be scored directly, showing recombination as it happens.

For most of the time, crosses between, say, an albino and a normally-coloured stock produce the expected 1 : 1 ratio of light and dark ascospores. Occasionally, though, something surprising happens: within an ascus there is an excess or a deficiency of one of the classes. Each ascus contains eight spores. Instead of the expected four of each class, there may be six mutants and two wild-type, or vice versa. One of the four copies of an allele in the meiosis (two on each of the chromosome pairs) has been converted into its alternative form.

Again, inexact recombination is to blame. There is a mismatch between base pairs as the homologous chromosome arms line up, distorting what should be a perfect fit. A repair enzyme cuts out the inappropriate pairings; if these are in the sections of DNA that determine the difference between mutant and wild-type, one allele can be converted into the other. If this process has its own impetus and acts directionally, favouring the conversion of one allele over another, the preferred form will replace its predecessor.

Another means of driving long sections of DNA towards homogeneity is the **transposition** of segments of the genome to new locations. Again, this is recombination (the formation of new genetic mixtures) albeit of a special kind. There are many ways in which this can happen. Some are based on DNA, but others involve reverse transcription, the transfer of information from RNA into DNA (Chapter 2). This is a powerful means of increasing the number of copies of a particular sequence by commandeering its product as a template. The first hint of this new evolutionary mechanism came from the breeding experiments on maize described in Chapter 4. P elements in *Drosophila* (Chapter 4) are also able to multiply themselves, and transposition may be responsible for some of the repetition in the genome.

All these mechanisms are alien to traditional Darwinism, but together they may mould much of the structure of the eukaryote genome. Sometimes, the process can be seen at work. One convincing example of how flexible the genome can be comes from primates. As we have seen (Chapter 4), the nervous degenerative illness Huntington's disease involves the expansion of a short repeated sequence, CAG, at one end of the gene. Once a critical number is reached, symptoms appear. The gene has inbuilt instability as the number of copies in a family with the disease may increase from one generation to the next as it passes through males.

Everyone has eight or more repeats – and this, of course, causes no symptoms at all. The most frequent number of repeats per individual is around 16. There is quite a lot of variation from person to person, with the distribution being quite strongly biased towards numbers above, rather than below, the average. In chimpanzees, our closest relative, the average number of repeats is nine – right at the bottom of the human range – and there is not much variation from chimp to chimp. Other primates, too, have low repeat numbers (Figure 5.14).

▷ What does the increase in number in the human lineage – but not our primate relatives – suggest about the evolution of CAG repeats?

▶ That it is a process of asymmetric recombination which began in the evolutionary branch leading to modern humans, and has never taken off in the other branches of the primate tree.

Perhaps, then, the increase in the CAG copies is a recent event in human evolution. There seems to have been a long-term mutational bias towards longer repeat numbers within the lineage leading to *Homo sapiens*, providing humans

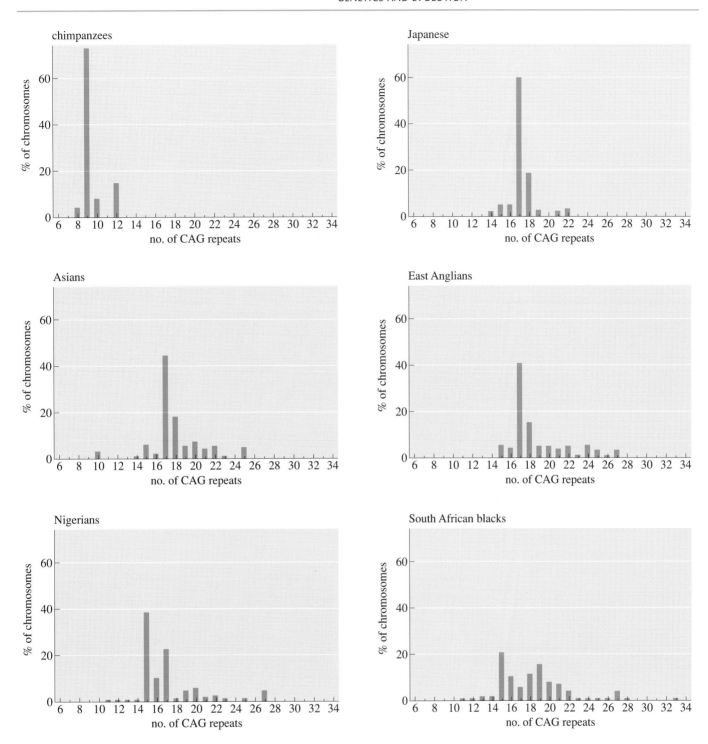

Figure 5.14 Population distributions of CAG repeats in the Huntington's disease gene in five human populations and in chimpanzees.

with more copies of this trinucleotide than any of their relatives and – in extreme cases – damaging a functional gene. Close to the CAG repeat is another region filled with triplet repeats, this time CCG. Once again, humans have many more copies of the repeat than do living primates, suggesting a general instability in this part of the human genome.

Sometimes, new cases of Huntington's disease appear in families that have never before experienced the disease. These new mutations arise by the expansion of the repeat number over the critical level of around 40 repeats. There are, in addition, great differences in the disease's incidence from place to place. In East

Anglia, one person in ten thousand has Huntington's, while only one Japanese in a million is affected. Remarkably enough, apparently unaffected people in East Anglia have, on the average, more copies of the CAG repeat than do their equivalents in Japan. The process of expansion is taking place more rapidly in East Anglia (and perhaps, indeed, the incidence of the disease will rise there in future generations). Why this is so, and why the human lineage has itself accumulated copies of this sequence, is not known: but it is an eloquent statement of just how unstable parts of the genome might be and how they appear to evolve of their own volition, without much reference to the world outside.

5.9 Evolution by genetic invasion

The camel, it is often said, is a horse designed by a committee. The structure of the genome is beginning to look like a similar uneasy compromise between conflicting biological needs.

It does seem, though, that nuclear genes, in general, confirm Darwin's idea – the longer ago that two creatures diverged, the more different they are. Most molecular clocks assume that evolutionary trees are, indeed, just like real trees: a common trunk which splits into smaller and smaller branches. However, certain large-scale patterns of relatedness suggest that there were some surprising events early in evolution. Whole genomes from one group of creatures have, apparently, been incorporated into completely different stocks. In the same way, a section of the genome with one function may have been hijacked by another evolving gene. Evolution may be like a tree; but sometimes branches can be grafted onto an alien trunk.

The grafts can take place between genes within an evolving lineage, or between different lineages. Certain functional genes are a pastiche of elements cobbled together from separate sources. In the rat growth hormone gene (a member of the prolactin–growth hormone gene family) there are five exons. Four share a similar sequence, suggesting that they themselves arose (as did the gene itself) by the duplication of an ancestor. The fifth exon is quite different in structure from the others and produces a peptide with a distinct function – reducing the activity of insulin. This exon probably invaded the growth hormone gene from somewhere else in the genome by what is known as **exon shuffling**.

Such movement of whole segments of genetic information among lineages is bound to confuse any evolutionary tree. Gene transfer may be a regular event, and may help to explain some of the more baffling attributes of the genome.

Genetic engineering began, in the eyes of genetic engineers, 20 years ago when it became possible to cut and splice DNA using restriction enzymes extracted from bacteria (Chapter 2). These were soon used to transfer genes between distantly related species, in contravention to what seemed the most fundamental rules of biology. The enzymes had, however, themselves evolved for a not dissimilar purpose; to cut out foreign DNA elements attempting to invade the bacterial genome. It now seems that such 'natural genetic engineering' – the movement of genes between distantly related species – may be generally important in evolution.

The genomes of many bacteria are full of alien genetic elements – plasmids, phages and transposable elements. Sometimes they are harmful; their persistence is no different from that of, say, tapeworms in humans. It depends on a balance between the harm the parasite causes and the rate at which it can spread. For much – perhaps most – of the time they do little damage (although the bacterium can get on perfectly well without them), and produce nothing which is expressed by the host.

In some circumstances, though, such extraneous genes are essential: they are closely associated with their host's survival and form an essential part of its genome. For a bacterium, life is full of nasty surprises, to which it must adapt or die. Very often its adaptation depends on genes imported from elsewhere – sometimes another species of bacterium – by a transposable element. Such genes include those for resistance to antibiotics, ultraviolet light and heavy metals; the ability to break down unusual food substances; the production of toxins that attack other bacteria; and the virulence of pathogens. The survival of bacteria depends on their ability to import genes from other lineages.

Much of this bacterial adaptation – itself a classic of Darwinian natural selection, however the genes might originate – has practical implications. It is common to find bacteria such as *Salmonella* in hospitals simultaneously resistant to five or more antibiotics. The genes for resistance have been picked up on plasmids over recent years. Ancestors of those plasmids stored away before the days of antibiotics do not carry the appropriate genes. The bacteria carrying the foreign DNA – and hence the plasmids themselves – have an advantage in the hospital environment, and quickly spread from species to species.

The process has gone so far in bacteria that foreign elements become incorporated into the host's genome. The phage λ, for example, can act as an infection, oozing out of its host's cells to attack another one (as described in Box 2.8 in Chapter 2); or be passed down the bacterial generations as the cells divide. When the phage is deprived of the chance to infect new hosts by keeping the bacteria at very low densities, it becomes less harmful to its host, in time becoming integrated into the bacterial chromosome and acting almost as a bacterial gene.

Multicellular organisms are also liable to invasion by alien DNA. You saw in Chapter 4 how many mutations are caused by disruption of a working gene by mobile DNA. P elements and other transposons have other properties that are difficult to fit into traditional evolutionary theories.

The various DNA-based elements all share a similar structure. They possess a number of short repeated DNA segments and at least one transcribed sequence which produces the transposase needed to copy them. The genus *Drosophila* possesses several distinct elements of this kind. Sometimes – as in the *mariner* species – it is possible to make an evolutionary tree based on DNA sequence change, showing the relationships between different elements. This tree suggests that the ancestor of *mariner* appeared at least 200 million years ago. Its own evolutionary tree shows only a weak association with the family tree of the insect hosts. The elements' promiscuous behaviour has led to many transfers between distantly related species.

▷ P elements are found in *Drosophila melanogaster*. Their closest relative is in a South American species of *Drosophila*. *D. melanogaster* spread from its African home – at the earliest – with Columbus. What does this suggest about the time in which P has resided within it?

▶ The interspecies transfer must have happened within the last few hundred *Drosophila* generations.

DNA-based transposable elements are found in groups as distantly related as maize, nematode worms and hagfish, suggesting that they have arisen on many occasions during evolution. Elements using RNA as a template are even more diverse. They all produce a reverse transcriptase, but are apart from that a rather heterogeneous group. Once again, related elements are found in very different hosts (the *copia* type in *Drosophila*, yeast and herring, for example), suggesting that they have moved among distinct lineages at some time during their history. Some related elements have lost the ability to produce their own reverse transcriptase and depend on an extraneous source of the enzymes. There may be thousands of copies of such short interspersed repetitive elements. One, the so-called *Alu* sequence, makes up a substantial part of the human genome (Chapter 2).

Transposable elements differ greatly in structure and degree of autonomy. Some, like P elements, hop frequently around the genome and move quite easily from one species to the next. Others, like *Alu*, are more lethargic, rarely proliferating and content to stay within one species. However, all are the products of evolutionary mechanisms different from anything imagined by Darwin. It is becoming clear that transposable elements have a major effect on the evolution of their hosts and explain some of the bizarre attributes of eukaryotic genomes.

The eukaryote cell itself may reflect a series of ancient evolutionary invasions of one lineage by another. As you will learn in the next chapter, the genes of mitochondria are more similar to those of bacteria than to the genes of the eukaryote nucleus, so that cooperation between very distant branches on the evolutionary tree may have been a feature of genetic change since soon after the origin of life.

5.10   Envoi: new answers or more questions?

The Origin of Species is, mainly, not about the origin of species. Instead it makes a convincing case that evolution has indeed taken place: that every living creature shares a common ancestor at some time in the more or less distant past. Darwin failed to answer some crucial questions, although he certainly recognized their importance. How biologically distinct are species and how do they arise? Why are there so many kinds of animal and plant, and why do they look so different? How could almost identical blobs of jelly – the fertilized egg of an elephant and a fruit-fly, say – develop in just a few cell divisions into animals that look completely unalike?

These questions remain largely unanswered today, although our ignorance is more detailed than it was in the time of Darwin. As we have seen in this chapter, genetics has set new questions about the structure of the genome which are hard to fit into the structure of conventional evolutionary theory.

However, what genetics has also done, with complete success, is to confirm the central tenet of *The Origin of Species*: that evolution – common descent – is a fact; one that gives coherence to the whole of biology. From the universality of the genetic code to the uncanny similarity of the base sequences of humans and chimps, modern genetics – in spite of its complexities – proves without doubt that all living creatures are relatives to a greater or lesser degree.

Darwin was fond of presenting what he called 'Grand Facts' at crucial points in his argument. Surely, for any biologist, this is the grandest fact of all.

Objectives for Chapter 5

After completing this chapter you should be able to:

5.1 Define and use, or recognize definitions and applications of, each of the terms printed in **bold** in the text.

5.2 Review the Darwinian model of evolution, place it into a Mendelian context, and discuss just what in modern genetics does and does not fit into conventional models of evolutionary change.

5.3 Discuss descent with modification and the use of phylogenetic trees based on genes to reconstruct the course of evolution using both viruses and higher organisms such as primates.

5.4 Show how advances in technology have changed our views of the amount of diversity present within species, and why this cannot easily be accommodated into the conventional view that natural selection is a universal force in evolution.

5.5 Produce a convincing case for the efficacy of selection in producing adaptive change, based both on laboratory experiments on ribozymes and on observations of genetic patterns in natural populations.

5.6 Discuss cases – both from human populations and in a more general context – in which striking genetic differences among populations are due, not to selection, but to accident; and address questions about the relative importance of selection and drift in evolution.

5.7 Assess the use of genes as 'molecular clocks'; how and why the rate of the clock may change with the molecule or the species involved, and the new insights which it has given into the date of origin of *Homo sapiens*.

5.8 Discuss what Darwinism can and cannot say about the structure of the genome; in particular by exploring the problems of repetition within species and rapid sequence divergence between them and analysing the probable importance of new genetical processes, including the transfer of genes among lineages, in leading to rapid changes in gene frequency.

5.9 Produce an argument that will convince every creationist that they are wrong.

Questions for Chapter 5

Question 5.1 *(Objectives 5.4 and 5.6)*

The children of cousin marriages are, on the average, about 4% less likely to survive to adulthood than those of unrelated parents. One estimate is that there are three million sites in the human genome at which there may be genetic differences between two randomly chosen individuals. What does this suggest about the possible action of natural selection at the molecular level?

Question 5.2 *(Objectives 5.2, 5.7 and 5.8)*

Chimps and humans share about 98% of their DNA; the two almost indistinguishable snails *Cepaea nemoralis* and *Cepaea hortensis* only about 80%. What does this suggest about their evolution?

CHAPTER 6

Inheritance outside the nucleus

6.1 Genetics with a single parent

This chapter deals with a part of the genome that once seemed unimportant: that located in the cytoplasmic organelles, in particular the mitochondrion. Because it is so small, the mitochondrial genome has been much studied. It has many unusual features. The DNA is arranged in a circle and in animals at least, it is economical, with none of the genetic redundancy found in the nucleus. It also has an unusual pattern of inheritance as mitochondria are largely passed down through the female line. Mitochondria are much involved in energy metabolism. Thus, mutations that damage mitochondria interfere with the energy balance of the cell. Metabolically active tissues – muscle or brain, for example – are particularly open to damage by mitochondrial mutation. Although only a few rare inherited diseases have been specifically associated with injury to the mitochondrial genome, other more common afflictions (diabetes, for example) can partly be blamed on injury to the energy-metabolizing machinery of mitochondria.

Many mitochondrial diseases arise from mutations within an individual's own body as cells divide; a hint that events in the soma may be as important as those in the germ line traditionally studied by geneticists. An increase in the incidence of mitochondrial mutations with age in somatic cells helps to explain the late onset of several diseases, and may be involved in ageing itself. Mitochondria lack repair enzymes, and evolve quickly. Consequently, they are often used to establish evolutionary phylogenies (including the human phylogeny claimed, misleadingly, to trace back to an ancient 'mitochondrial Eve'). Their own evolutionary history is remarkable. An evolutionary tree based on slowly-evolving DNA sequences unites mitochondria with bacteria, rather than with genes of the eukaryote nucleus. So, mitochondria, chloroplasts and perhaps other cell organelles probably originated as intracellular parasites: and, in the 2 000 million years since then, have lost many of their genes to the nucleus, gaining a few in return.

Genetics is, conventionally, about sex. Genes are passed to offspring through males and females. With some differences in detail, each sex makes an equal contribution to the next generation. Because of recombination between male and female genomes new mixtures of genes appear each generation.

There are, however, huge areas of genetics which do not depend on sexual reproduction at all. Many organisms – from potatoes to certain lizards – consist only of females.

▷ What are the advantages of asexual over sexual reproduction?

▶ Asexual reproduction is a more efficient way of multiplying an individual's genes. A parthenogenetic female can produce twice as many replicas of her own genome as would be the case if she had to dilute her reproductive efforts by receiving and replicating copies of genes from a male. Why sex evolved is thus far from clear.

So far, so simple. But what is surprising is to learn that eukaryotes have within themselves both sexual and asexual methods of passing on different genes.

Early in the history of modern genetics, certain unusual patterns of inheritance suggested that there must be genes outside the nucleus. Such **extranuclear genomes** have since been found in many species.

The first hint that there is more to reproduction than sex came from crosses made on plants. In 1909 – less than a decade after the rediscovery of Mendel's Laws – Charles Correns was studying the inheritance of various characters in the four o'clock plant, *Mirabilis jalapa* (Figure 6.1). As is true for many decorative plants, some of its branches may bear green leaves, some white. Other branches may be variegated, with green and white patches on the leaves. Flowers form on each branch, and can be used for crosses. The genotype of the gametes produced by the flower is the same as that of the branch upon which they grow.

Many *Mirabilis* variants are inherited in a straightforward Mendelian way. But there is one oddity. When pollen from flowers on green branches (which had normally-pigmented leaves) was used to fertilize the eggs of branches whose leaves were white, all the offspring were white. However, the opposite cross – pollen from white with eggs of green – gave a different result. All the offspring were green. The same results appeared in a subsequent generation when white and green branches were crossed together: the offspring always resembled the mother, so that this is not a simple case of sex-linkage. Instead, the pattern of inheritance is **matrilineal**, only mothers passing on the character although both daughters and sons receive it.

Some of the plants had variegated leaves, with patches of green and white. Pollen from plants with such leaves used to fertilize 'green' or 'white' flowers again produced offspring resembling the mother. However, the reciprocal cross gave an odd result. Many of the offspring were variegated like their mothers, but a few were green and a few white. Crosses with these green and white plants gave the usual pattern of straightforward matrilineal inheritance as before; but crosses using the variegated plants as mothers gave, once again, green, white and variegated plants in the offspring.

What was going on? Clearly, this was not Mendelian inheritance: there was not a simple segregation of distinct phenotypes each generation. The whole business with variegation looked at first sight alarmingly like the blending inheritance which had caused so much trouble in the early days of genetics.

In fact, something more conventional is happening. Males and females differ in one attribute: the size of their germ cells: pollen (or, in animals, sperm) and egg. Sperm and pollen are much smaller than eggs. The variety of sexual experience in the living world is so enormous and the difference in appearance of the sexes so great that this **anisogamy**, as it is known, is the best method of defining what males and females actually are. Generally speaking, the male gamete passes on only the contents of its nucleus (including, of course, the nuclear DNA), while eggs pass on not only their nucleus but also the cytoplasm which surrounds it.

Cytoplasm consists of all kinds of structures. Some are under the direct control of genes in the cell nucleus. Others, though, are more autonomous. Some cytoplasmic elements are parasites or symbionts. They live within cells, doing not much damage, and are passed on to the next generation through the egg cytoplasm. Others play an essential role in the activities of the cell.

The green colour of plants comes from chlorophyll, held in specialized cytoplasmic organelles, the chloroplasts. In white or variegated four o'clock plants the

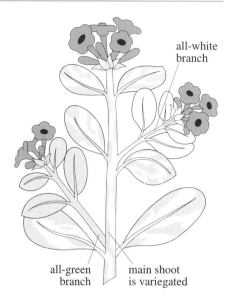

Figure 6.1 Leaf variegation in the four o'clock plant, *Mirabilis jalapa*. Flowers may form on any branch (green, white or variegated) and these flowers may be used in crosses.

chlorophyll-synthesizing machinery has been damaged by mutation. When a white male was crossed with a green female, all the offspring's chloroplasts came from the mother's cytoplasm, and were – like hers – normal. In the reciprocal cross the offspring again resembled the mother and were white. Variegated mothers had both kinds of chloroplast. When the egg was formed, some gained only white, some only green, and some both genetically distinct types of chloroplast.

▷ In crosses with a variegated mother, the pattern of variegation – the size of the green and white patches – varies between individual offspring. Why might this be?

▶ The size of green or white patches in a variegated plant depends on the proportion of the two chloroplast types segregating as the egg cytoplasm splits. Within the growing plant, some cell lineages receive an excess of 'green' and some an excess of 'white' chloroplasts. The balance between the two determines the size of the patches.

Genetic information can, Correns suggested, be passed down via the cytoplasm as easily as through the nucleus. The cytoplasm has genes in animals, too. Here the most important cytoplasmic genome is that of the mitochondria: cellular organelles with DNA of their own which is passed down the female line in the egg cytoplasm. Because it is small and can easily be isolated, mitochondrial DNA is now the best-known section of the eukaryotic genome. The structure, organization and evolution of mitochondrial genes are much better understood than are those of the nucleus. In addition, mutations in mitochondrial DNA are known to cause a variety of inherited diseases.

6.2   Mitochondria: an overview

Mitochondria are found in the cytoplasm of all eukaryotic cells (Book 2, Chapter 4). The are about 1–$3\,\mu m$ long – the size of bacteria – with a smooth outer membrane and an inner one folded into many compartments or **cristae** which project into the matrix. On the surface of the membrane and embedded within it is the protein machinery for aerobic respiration.

▷ Some fungi (*Neurospora*, for example) have long been known to possess matrilineally-inherited mutations that slow the rate of growth. What does this suggest about their genomic location?

▶ Such mutations are in the mitochondrial DNA, where they result in interference with energy production and developmental retardation. This was, in fact, the first hint that mitochondria had mutations of their own.

As you learnt in Book 2, mitochondria are concerned with energy conversion: they use energy from aerobic respiration to phosphorylate ADP to ATP. The enzymatic addition of a single additional phosphate group to ADP traps the chemical energy released by the oxidation of sugars and fats (and in plants by photosynthesis). Hydrolysis of ATP and the loss of a phosphate produces ADP, releasing energy in the process. As we shall see, many of the inherited errors in mitochondrial DNA produce symptoms that can be related to faults in the energy-conversion machinery.

Mitochondria often divide. The rate of division is regulated by the nucleus in response to the metabolic needs of the cell. Active cells have more mitochondria.

Liver cells, for example, may contain up to 3 000, about a quarter of the cell volume. Heart muscle cells have even more, but less active cells (such as those in epithelia) contain far fewer. Mitochondria also have a certain developmental flexibility and can vary in size. Those in muscle cells have more cristae, providing a greater surface area for aerobic respiration.

6.3 The chloroplast and mitochondrial genomes

The mitochondrial genome codes for only a small proportion of the macromolecules found within the organelle. In humans it consists of a single circular molecule of DNA. The fact that organelle genomes were indeed circular was discovered long before the advent of modern technology. In this, the history of organelle genetics resembles that of the study of nuclear genes: the major features of the map were inferred from ingenious crossing experiments long before chemistry came to the rescue and allowed its structure to be determined directly.

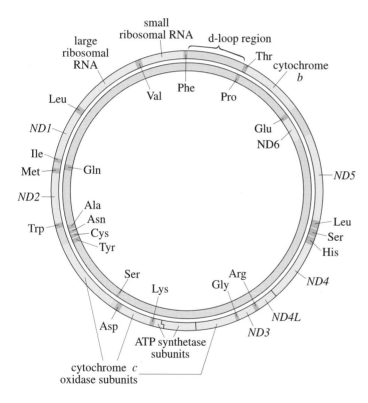

Figure 6.2 Organization of the human mitochondrial genome.

▷ In the green alga *Chlamydomonas*, chloroplasts carrying different mutations occasionally find themselves within the same cell and there is recombination between their genomes. The linkage map built up from recombinant frequencies has an odd property. The order of loci changed from experiment to experiment. Three loci might be mapped in the order A-B-C in one cross; only to appear as B-C-A in the next. How can this be explained?

▶ The only way to resolve the paradox is to conclude that the map must be circular, and that the breakage and recombination take place at different places in the circle in different experiments. This was the first hint that the chloroplast genome is not (as is the case for nuclear genes) arranged in a straight line.

There may be as many as ten copies of the circular genome within a single mitochondrion. New mitochondria are produced by simple fission, in much the same way as bacteria duplicate. However, mitochondrial reproduction is not autonomous: most of the protein components needed are coded for by nuclear genes, translated into protein by cytoplasmic ribosomes, and then transported from the cytoplasm into the organelle. This process involves several distinct mechanisms; including the recognition of specific amino acid sequences by components of the outer mitochondrial membrane and the activity of chaperone proteins that form complexes with and accompany newly synthesized proteins (Book 1, Chapter 4).

There is extraordinary diversity in the structure of the mitochondrial genome in different creatures. Its size varies by nearly 200-fold, from only 13.8 kbp (kilo base pairs) in the nematode worm *Caenorhabditis* to 2 400 kbp in the muskmelon *Cucurbis*. Despite the size difference, the respiratory chain genes in different species are very similar and the prime role of mitochondrial genes remains to code for components of the electron transport system.

Size variation in mitochondrial DNA is largely a reflection of differences in its organization. Mammalian mitochondrial DNA (Figure 6.3) has very high information density: coding regions abut directly, and sometimes even overlap (with the first nucleotide of one gene acting as the last nucleotide of the preceding gene). There are very few introns and over 90% of the genome is transcribed. Human mitochondrial DNA is about 16.5 kbp long. There are 37 coding genes, specifying 22 tRNAs, 13 mRNAs (coding for polypeptides that form part of the electron transport chain) and two rRNAs (coding for RNAs of the small and large subunits of the ribosome). The non-coding *control region* is between 1 and 2 kbp long.

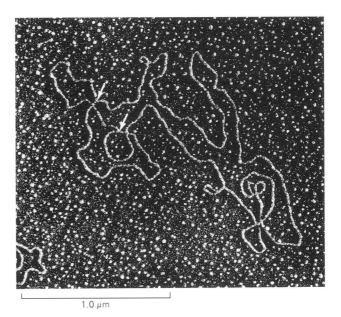

1.0 μm

Figure 6.3 Electron micrograph of a mammalian mitochondrial DNA molecule caught during the process of DNA replication. The circular DNA genome has only replicated between the two points marked between the white arrows.

Animal mitochondrial DNA, whatever its size, has a relatively stable gene order. Humans, mice, rats, cows and the toad *Xenopus* have exactly the same order of mitochondrial genes, while that of the chicken has a translocation that changes it slightly. Sea urchin and *Drosophila* mitochondrial DNA differ from that of vertebrates by two translocations and three inversions (Chapter 4), respectively. Not all mitochondrial genomes are as conservative, though: the gene order in nematodes is very different from that of other animals.

Flowering plant mitochondrial DNAs are large, ranging from 200 to 2 400 kbp Comparison of related species shows that their mitochondrial DNA has a propensity for rapid structural change because of frequent recombination. Plant mitochondrial DNA has only a few genes in addition to those found in animals. In striking contrast with mammalian DNA, though, only about a tenth has a coding function. There are many introns and repeated sequences and the mitochondrial genome has also recruited sequences originating from chloroplast DNA and from the nucleus during its evolution. The function of such 'promiscuous' DNA, if indeed it has any, is not known.

Some mitochondrial DNAs are best described as bizarre. That of *Trypanosoma* is a network of thousands of interlocked DNA circles. Some of these, the *maxicircles*, have a compact genome and do the same job as the mitochondrial genomes of other animals. What the thousands of *minicircles* do is not known.

6.4 Transcription, translation and replication of mitochondrial genes

Mitochondrial genes are transcribed by an RNA polymerase coded for within the nucleus. In mammals, the transcriptional control signals are in the **displacement- or d-loop**. The mRNAs transcribed from mitochondrial genes are translated and expressed within the mitochondrion itself. Remarkably enough, the mitochondria of many species show exceptions to the 'universal' genetic code used in the nucleus, particularly in codons involved in starting or stopping protein synthesis. It might be that the mitochondrial pattern is more primitive; but, as the changes in code differ from species to species they may have evolved in response to the special needs of protein synthesis in the mitochondrion.

Mitochondrial DNA replication starts in the d-loop, which is 500–600 bp long in mammals and contains the *origin of replication* of the L (light) strand. Figure 6.4 shows how the H (heavy) strand is used as a template for the synthesis of another L strand. The d-loop is extended, displacing the original L strand as far as a point about two-thirds of the way around the circle, where an origin of replication on the H strand is exposed. Synthesis of the complementary strand begins here and proceeds in the opposite direction. In contrast to the RNA polymerase, mitochondria have their own DNA polymerase, different from that in the nucleus.

▷ What do the unique biochemical features of the mitochondrial code and replication mechanisms suggest about the history of the organelle?

▶ That it may originate in a different evolutionary lineage from that giving rise to nuclear genes.

6.5 Mitochondrial inheritance in humans

We receive our mitochondria from the cytoplasm of our mothers' eggs. Sperm have mitochondria, but these are in the *midpiece*, which is left behind on fertilization. (Not all animals show this neat pattern: in mice there is a small amount of 'leakage' of mitochondrial genes through the male line, and in some molluscs sperm pass on large numbers of mitochondria.)

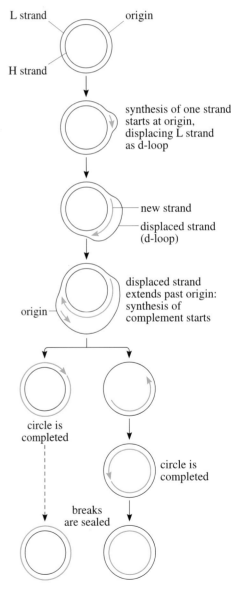

Figure 6.4 Replication of mitochondrial DNA.

▷ What is the characteristic feature of pedigrees of diseases caused by mitochondrial mutations in humans?

▶ Both sons and daughters receive their mitochondrial DNA (and thus the gene causing the disease) from their mothers, but only daughters pass it on to the next generation.

Figure 6.5 shows the pattern of inheritance of a neuromuscular disease caused by a mitochondrial mutation. In practice, the symptoms of diseases arising from mitochondrial mutations depend on the tissues receiving the damaged mitochondrion and on age, but this matrilineal inheritance is a good clue that a pedigree may indeed represent a mitochondrial mutation.

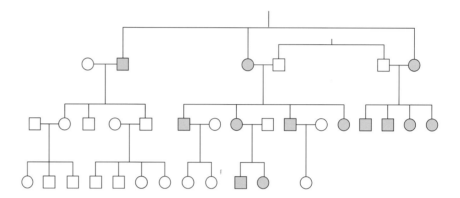

Figure 6.5 Pattern of inheritance of a neuromuscular disease caused by a mitochondrial mutation.

▷ How does matrilineal inheritance differ from sex-linked inheritance?

▶ In most crosses involving recessive X-linked inheritance only males are affected in the next generation: heterozygous daughters pass on the allele without manifesting its effects. The mutation can also descend through males.

Mitochondrial mutations may be hard to identify. Each cell contains many mitochondria, and each mitochondrion has several DNA molecules. Thus, a mutation may be present in only some of the mitochondria passed on to the next generation; and, within a cell, there may be *heteroplasmy*, the presence of both mutant and non-mutant mitochondrial genomes. During cell division, the proportion of the two genomes may change because of the segregation of cytoplasm, and hence mitochondria, during mitosis and through variation in the rate at which mutant and normal mitochondria divide. Cells may then drift towards *homoplasmy*; all mutant or all normal mitochondrial genomes. This, in turn, may have effects on the phenotype of a particular tissue.

Many diseases that result from mitochondrial mutations are due to abnormal proteins involved in respiratory chain reactions.

▷ What types of tissue are expected to be particularly susceptible to the damaging effects of a mutated mitochondrion?

▶ The importance of oxidative phosphorylation varies from tissue to tissue. Consequently, mitochondrial defects hence do their most serious damage to metabolically active tissues such as brain, retina, heart and muscle.

Several inherited diseases are now known to result specifically from mitochondrial mutations. As is the case in nuclear genes, these may arise from point mutation or from deletions and insertions. (See Figure 6.6.)

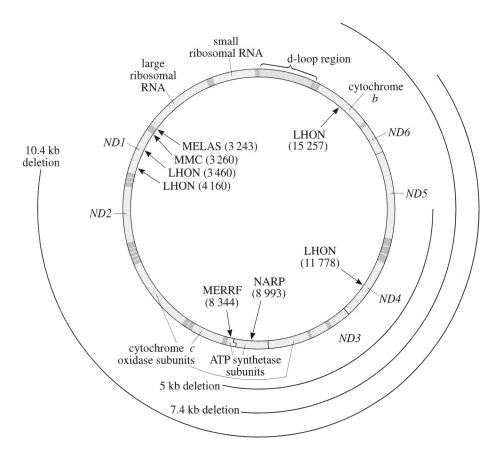

Figure 6.6 Some mitochondrial deletions and point mutations. MELAS, mitochondrial encephalomyopathy, lactic acidosis and stroke-like symptoms; MMC, maternally inherited myopathy and cardiomyopathy; LHON, Leber's hereditary optic neuropathy; MERRF, myoclonic epilepsy and ragged red-fibre disease; NARP, neurogenic muscle weakness, ataxia and retinitis pigmentosum. Numbers are distances in base pairs from the origin of replication.

Leber's hereditary optic neuropathy (LHON) is a maternally inherited form of blindness. Beginning in a sufferer's mid-twenties, the optic nerve begins to degenerate. Several distinct mitochondrial mutations are associated with the disease (four of these are shown in Figure 6.6). Their severity varies with the particular respiratory enzyme damaged; and there are hints that a gene in the nuclear DNA, perhaps on the X chromosome, may also be involved.

There are several other mitochondrial point mutations that can cause disease. Often, heteroplasmy affects the symptoms. In one group of mitochondrial diseases, the extent of damage depends on the proportion of mutant mitochondrial DNAs within a particular cell lineage. In one family, several individuals were affected. One child was so badly damaged that more than 95% of the mitochondrial genomes were abnormal within the fibroblasts, brain, kidney and liver. She was diagnosed as having Leigh syndrome, a severe degenerative brain disease and died when only 6 months old. Her mother, aunt and uncle, who had all inherited the same mutation from her grandmother, all had different severities of disease. Her maternal aunt had the same proportion of abnormal mitochondrial DNA in her fibroblasts and died at the age of one year. Her maternal uncle, though, had about 80% abnormal mitochondria in his tissues, and appeared to be normal until he was about 12. He then suffered from muscle weakness, followed by mental deterioration and an atrophy of the retina leading to blindness. The mother had 61% abnormal mitochondrial DNA in skin fibroblasts and 39% in lymphocytes and was phenotypically normal. It is clear then that, in this pedigree, the proportion of abnormal mitochondria within particular tissues has a critical effect on the fate of those inheriting them.

Other base substitutions affect protein synthesis in the mitochondria. Many involve mutations in a tRNA. Again, heteroplasmy is important, and symptoms only manifest themselves as the proportion of abnormal mitochondria rises above 85%. A mutation in one particular mitochondrial tRNA gene causes myoclonic epilepsy (whose symptoms include jerky movements and mental deterioration) and ragged-red fibre disease – which distorts the appearance and function of muscle fibres and sometimes produces kidney failure and deterioration of the heart. This is but one of a whole family of muscle diseases known as the mitochondrial myopathies.

Deletions in mitochondrial DNA may be small or quite substantial (Figure 6.6). Some – usually involving between 2 and 6 kbp – lead to diseases of the eye muscles (ocular myopathies). A larger deletion that destroys the ability of mitochondrial DNA to replicate leads to deafness and sometimes to diabetes; and yet another damages the cells that produce blood cells. Duplications of short sections of the mitochondrial genome produce symptoms superficially similar to those caused by small deletions.

6.6   Mitochondrial mutations and age

Because the overall efficiency of the respiratory chain decreases with age and because mitochondria accumulate mutations as they divide, some mitochondrial diseases show their symptoms only in older people. As we saw in Chapter 4 (Section 4.9), the rate of germ-line mutation goes up dramatically with age. The same is true of mutation in somatic cells: a theme that is explored further in the next chapter of this book. Indeed, some theories of ageing are based on this accumulation of mutations in older cells that have undergone more cell divisions.

The effect is certainly seen in cytoplasmic organelles. Mitochondrial genomes from the tissues of elderly people have suffered more mutational damage (often involving quite large deletions) than have those from younger individuals. This may explain why some inherited mitochondrial diseases – whose mutations are, of course, present at birth – manifest their effects only in middle age, i.e. when additional DNA damage has reduced the numbers of normally functioning mitochondria. Figure 6.7 shows the accumulation of a particular mitochondrial mutation in heart tissue from patients with coronary heart disease. The damage to the mitochondria means that patients have a chronic oxidative phosphorylation shortage – and, if they live long enough, ATP production in the heart (which is a metabolically demanding organ) falls below the level needed for normal function.

Patients with certain mitochondrial diseases such as ocular myopathy (paralysis of the eyes, loss of hearing, heart defects and dementia) or Pearson syndrome (loss of blood cells due to the failure of bone marrow precursor cells) all have a mitochondrial deletion absent from other members of their family.

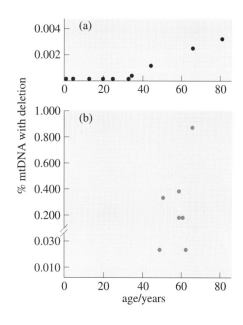

Figure 6.7   The percentage of cardiac mitochondrial DNAs with a particular deletion in (a) normal hearts (black circles) and (b) hearts of individuals with coronary heart disease (coloured circles), as a function of age.

▷ What does this suggest?

▶ That the syndromes are due to new mutations in the body cells of the patient. It also indicates that transmission of large deleted mitochondrial DNA through the germ-line is very limited, perhaps because such mutations are lethal early in development.

There is a great diversity of symptoms in certain mitochondrial diseases. People with ocular myopathy have one or more of a wide variety of problems, including hearing loss, uncontrolled shaking, heart defects, ptosis (drooping of the upper eye lid due to loss of control of the third cranial nerve) or paralysis of the muscles of the eye.

▷ What can be inferred from the variation in severity of symptoms from tissue to tissue in different patients?

▶ The mutation arose in different somatic tissues in different patients or, if the mutation arose early in the developing embryo, different tissues have different proportions of the mitochondria with deleted genomes.

Another reason for the worsening of symptoms with age is that the proportion of mitochondria with the deletion increases. The rate of multiplication of any DNA molecule depends on its length. As deleted and normal mitochondrial DNA molecules turn over, those with the deletion replicate faster and increase in number. This, in turn, exacerbates the symptoms of the disease.

Mitochondrial mutations in somatic tissue may even help to understand why many diseases involving damage to metabolism (such as adult-onset diabetes) become more destructive with age. Parkinson's disease, too, is associated with deletions of mitochondrial DNA in the tissues of the brain (although this is not the primary cause of the damage). Mitochondria may have a more fundamental role in disease, ageing and bodily decay than has previously been imagined.

6.7   Rates of mitochondrial DNA sequence divergence

Mammalian mitochondrial DNA *structure* evolves slowly: gene order does not change. However, *sequence* divergence among vertebrates is rapid, with up to ten times as many differences among species than in equivalent sections of coding DNA in the nucleus. Part of this is due to the inefficiency of the mutation repair mechanisms (Chapter 4, Section 4.8) in mitochondrial DNA and with the decreased fidelity of DNA replication. Mammalian mitochondrial DNA is the most rapidly evolving genome yet discovered.

Oddly enough, invertebrates do not have a particularly rapid rate of mitochondrial DNA sequence divergence. Sea urchin or *Drosophila* mitochondrial DNA evolves at a rate similar to that of nuclear DNA. Plants show the opposite pattern to that of vertebrates. In spite of its rapid structural evolution through genome reorganization, plant mitochondrial DNA sequence evolves ten times more slowly than that in the nucleus – in fact, the plant mitochondrial genome has the slowest evolving sequence of any yet discovered. The reason for these differences among taxonomic groups is not known.

The rapid rate of nucleotide sequence evolution means that vertebrate mitochondrial DNA has been extensively used to study vertebrate evolution.

It contains within itself the history of females. Many evolutionary trees showing the relatedness, inferred from sequence data, of different mitochondrial lineages have been produced. Often, there is more geographic subdivision for mitochondrial genes than for those in the nucleus. This is true for African pygmies, for example: mitochondrial genotypes are more localized than are those passing through both males and females. Comparing nuclear and mitochondrial DNA patterning shows that males move, on average, more than females.

For humans, a world-wide family tree showing the relatedness of mitochondrial genotypes produced a genealogy leading back to a single woman (inevitably named 'mitochondrial Eve') who lived, it has been claimed, in Africa around 200 000 years ago. Of course, in the absence of recombination, any set of homologous genes must descend from a common ancestor (although each gene set will trace back to a different ancestor). Surnames are an approximate analogy. They descend through only one sex (the males in this case). Everyone with the same name (at least with a name that originated only once) can trace their descent from the same distant ancestor, the first to bear that name. In some places, there are rather few surnames. The existence of perhaps 20 names in a particular community does not, of course, prove that there was a time in its history when there were only 20 men present. The surname of any man who fails to have a son (either because he just has daughters, or has no children at all) will disappear. In just the same way there were many women living at the same time as 'mitochondrial Eve'. She is interesting only in that her mitochondrial genes are the only ones to survive the statistical probabilities of extinction. Whether 'Eve' was an African – or whether, in any real sense, she existed at all – is not clear.

6.8   The evolution of mitochondria: cells as chimaeras

How did mitochondria arise? Was it by the subdivision of a single cell into distinct compartments (as happened at the origin of the nucleus) or by the fusion of formerly free-living prokaryotes? The idea that a new creature could appear from the fusion of separate organisms goes back to the Greeks. The Chimaera had the head of a lion, the body of a goat and the tail of the dragon; and the Harpy was a monster with the face of an old hag, the body of a bird and the ears of a bear. Such beasts gained their fabulous power because the concept that the union of different creatures can produce viable offspring is so alien to human experience. Darwin, no doubt, would agree: his view of evolution involved the unfolding of one lineage to produce a diverse set of descendants. As we saw in Chapter 5, there are several instances in which short segments of DNA have invaded one species from another, sometimes bringing genes that are of value to the recipient. Now it seems that, early in the history of life, there were some larger-scale transfers of genes among very distinct evolutionary lineages.

Composite organisms began to be taken seriously by biologists with the discovery that a lichen is a compound plant consisting of a fungus and an alga; a relationship known as **symbiosis**. Both organisms benefit equally from this relationship. **Symbiogenesis** describes the notion that this process could go further and lead to the permanent fusion of two distantly related organisms. It leads to the genetic invasion of one evolving group by another.

There are many intracellular associations of eukaryotes and specialized prokaryotes. Some of the invaders are no more than parasites. Others, though, develop permanent and mutually dependent relationships with their hosts and are transmitted in the cytoplasm. The ciliated protistan *Paramecium bursaria* harbours eukaryote green algae that rely on the *Paramecium* for their survival and in return provide the host with the products of photosynthesis. *P. aurelia* carries various bacteria in its cytoplasm. They have a less permanent position, as their maintenance – or ejection – depends on genes in the host. Some of these bacteria, the so-called kappa particles, produce a toxin which is secreted by the host *Paramecium* and kills other *Paramecium* with a different genetic constitution. The symbionts and some (but not all) of their hosts have come to a genetic arrangement that allows one to survive within the other.

There are many structural similarities between mitochondria and bacteria, and it is now generally accepted that the organelles originated from a bacterial endosymbiont early in the history of the eukaryotic cell. The most convincing evidence comes from molecular biology.

The structure of mitochondrial DNA is more like that of a bacterium than that of the eukaryotic nucleus. Both are closed circles and lack the histone proteins associated with nuclear DNA. In mammals, there are no introns in the mitochondrial genome – and the same is true of most bacterial genomes. Some of the triplet codons (including the 'start' and 'stop' signals) differ from those of the nucleus, and mitochondrial DNA – like that of prokaryotes – is never methylated (Chapter 3). In addition, mitochondrial ribosomes resemble those of bacteria and are smaller than the cytoplasmic ribosomes. Certain poisons inhibit protein synthesis in both mitochondria and bacteria, but not in the cytoplasm.

All this might, of course, be convergent evolution (structural or functional analogy) rather than common descent (homology). DNA sequences, though, strongly support the endosymbiotic theory. For example, the cytochrome oxidase genes of the purple bacterium *Paracoccus denitrificans* are very similar to those of mitochondria. More remarkably, some *nuclear* genes coding for proteins synthesized via the mitochondrial transcription and translation apparatus are similar in sequence to their equivalents in *E. coli*, suggesting that certain genes of bacterial origin have been shifted to the nucleus since the invasion of the precursors of mitochondria early in cellular evolution.

The strongest evidence that nuclear and mitochondrial genomes have a separate origin comes from studies of the evolutionary relatedness of their DNA. This is best manifest in a molecular tree based on changes in the DNA sequence of the small ribosomal RNAs. These genes evolve very slowly, retaining a strong hint of their earliest ancestry. The tree shows that bacteria represent two distinct lines of descent (the Archaebacteria and the Eubacteria) while eukaryotes exemplify a third. All three lineages have small ribosomal RNAs; and their sequences can be aligned by comparing a highly conserved segment at the heart of the molecule – the u*niversal core*.

The tree unites mitochondria with purple bacteria (Proteobacteria), rather than with eukaryotic nuclear DNA. Members of this group include *Rickettsia*, Agrobacteria and Rhizobacteria. Many of these have intracellular associations with eukaryotic cells (for example, the nitrogen-fixing *Rhizobium* in root nodules of legumes). Analogous genes within plastids (such as chloroplasts, the site of photosynthesis) show that these organelles evolved from the blue–green algae (Cyanobacteria).

Some purple bacteria have properties that hint at what may have happened during endosymbiosis. For example, *Agrobacterium tumefaciens* causes crown gall, a tumour of plants. It transfers its own DNA to the plant genome, where it integrates and is multiplied by the plant's own replication machinery. Perhaps these organisms have an intrinsic ability to invade and form alliances with distantly related ones.

Although to our eyes bacteria as a whole appear rather uniform, their DNA sequences show that as a group they are much older and more diverse than is the whole of the animal kingdom. Mitochondria and plastids hence originate in very distinct divisions of the living world. It is not certain whether all mitochondria (or all plastids) have a single common ancestor or whether there were several invasions of the eukaryote lineage. The mitochondria of some green algae do appear to be sufficiently different from those of other plants and of animals to suggest, if not a completely separate origin, at least a very early split in their evolutionary history.

6.9   Evolution since the invasion: trading genes between cytoplasm and nucleus

As we have discussed, modern mitochondrial and chloroplast genomes resemble, but are not identical to, those of bacteria. The ancestral genome has been much modified since it joined the cell. It has been reduced and reorganized, and seems to have lost many of the genes present in its bacterial relatives.

▷   In maize, most of the mutations that produce chlorophyll-free albino plants follow the straightforward rules of Mendelian segregation, even though they exert their effect on the chloroplast. Why should this be?

▶   The maize chloroplast genome has shipped off many of its genes to the nucleus, where they lost their matrilineal pattern of inheritance although the phenotypic effects of mutations may be the same as those in the chloroplast itself.

The functional part of the mitochondrial genome, too, is smaller than that of any bacterium. Soon after the invasion the 'proto-mitochondrion' must have contained all the genes it needed; but most of the information needed to make mitochondria now resides in the nucleus. Some of the mitochondrial genes may have become redundant as they were already present in the host, but others have certainly been transferred from the ancient mitochondrial ancestor to the host's nuclear genome. Some pathways – including those responsible for ATP and NADH metabolism – involve enzymes partly coded for in the nucleus and partly in the mitochondrion. Sometimes the process is incomplete: in yeast one of the genes coding for an ATP synthetase subunit remains in the mitochondrion whereas it has been transferred to the nucleus in mammals.

Occasionally it is possible to guess at precisely what happened during the course of mitochondrial evolution. One mitochondrial gene in flowering plants codes for a particular subunit in the respiratory machinery. In most angiosperms it is coded for only by mitochondrial DNA, but in the cowpea it is coded for in the nucleus. In soya beans there are two copies: an active one in the nucleus and a copy that is never expressed in the mitochondrion. In both cowpeas and soya beans, the nuclear version of the gene is more similar to an edited transcript of the gene than to the unedited mitochondrial gene sequence.

▷ What does the lack of introns in the nuclear copy of this plant gene suggest about how it was transferred from the mitochondrion?

▶ That the process involved reverse transcription of an edited mRNA intermediate.

Why there should be this trading of genes is not certain. It might reflect natural selection among mitochondrial genomes. Like their bacterial ancestors, their main mode of reproduction is asexual and – like them – they are in constant competition with neighbours eager to pass on their own genes. Slimming down the genome by cutting out introns or shipping genes off to the nucleus speeds up the rate at which mitochondrial DNA can be replicated. Bacteria bearing plasmids replicate more slowly than those lacking this extra DNA and, as we have seen, there is evidence from human disease that mitochondria with deletions may gradually take over a growing tissue because of their greater speed of replication. This explanation is plausible for vertebrates, but is less satisfactory in explaining the enormous mitochondrial genomes found in plants.

Some theories of the origin of the eukaryote cell suggest that it is, effectively, an ancient alliance of several disparate ancestors. Some cell organelles (including mitochondria and chloroplasts) retain some of their DNA while others may once have been autonomous but have now shipped off all their genes to the nucleus, leaving no evidence of past independence.

The autonomous genome which remains within the mitochondria has, as we have seen, a genetic system of its own. Many of the processes involved take place in body cells (the **soma**) rather than the germ line. Geneticists are only now beginning to realize that such somatic processes are as important to nuclear genes as they are to those in the cytoplasm. The study of *somatic cell genetics* epitomizes the power of the new genetics in helping to understand the workings of the body. It is discussed in the next chapter.

Objectives for Chapter 6

After completing this chapter, you should be able to:

6.1   Define and use, or recognize definitions and applications of, each of the terms printed in **bold** in the text.

6.2   Produce a pedigree showing matrilineal inheritance, discuss its relevance to organelle (chloroplast and mitochondrion) genetics, and address the relative importance of organelle mutations in the germ line and in somatic cells in leading to disease.

6.3   Review the function of the mitochondrion, its structure and its mode of division.

6.4   Outline the major features of mitochondrial DNA, including its circular form, its gene arrangement and its high mutation rate.

6.5   Illustrate organelle inheritance with reference to human mitochondrial diseases; and understand why the symptoms of such diseases are tissue-specific and often change with age.

6.6   Discuss the use – and abuse – of mitochondrial family trees, and how these inevitably lead to a single ancient matriarch.

6.7   Review the evidence that cell organelles probably descend from an ancient invasion of eukaryotes by a bacterial symbiont.

Questions for Chapter 6

Question 6.1   *(Objective 6.2, 6.5, 6.6)*

What separates sex-linked inheritance from mitochondrial inheritance?

Question 6.2   *(Objective 6.5)*

Name three features that would, in the absence of detailed pedigree data, suggest that a particular syndrome was due to a mitochondrial mutation.

Somatic cell genetics

7.1 Introduction

This book has been a tale of the germ line – of the genetic processes associated with sex. The whole of evolution depends on variation and diversity, and the transmission of a dynamic genome from generation to generation. But the processes of genetics are not confined to germ cells. Genetic change also occurs in *somatic* cells. A population of cells (or, for that matter, of individuals) without sexual recombination – that is, gene recombination due to independent assortment and to chiasmata formation (crossing over) – is in danger of accumulating harmful mutations. Each new mutation in any asexual lineage that escapes intracellular repair (Chapter 4), is then subject to the stringent test of natural selection.

Sex may even have evolved as a means of removing the damage which accumulates through mutation as cells divide. It separates the fate of mutations from that of those who carry them as, in each generation, it produces new recombinant genotypes, that is, new mixtures of genes. Some individuals receive, by chance, several harmful mutations, others none. Those unlucky enough to inherit certain combinations of mutations do not survive; removing, with a single death, several damaged genes. Others thrive, passing on a genome purged of damage.

How does somatic change affect the phenotype? What are the consequences for the individual? Are changes in the genome at the somatic level harmful or harmless? Like germ-line mutations, do somatic mutations increase with age? Do they contribute to the survival of the organism or not? These are the questions we explore in this chapter. Somatic change – like germinal mutation in the early days of genetics – is sometimes seen as a wholly harmful process. In fact this is far from the truth. The immune system, crucial to the body's defences, *depends* upon it. However, the negative side of somatic change is seen in cancer, now known to be largely a genetic disease of somatic cells. We illustrate the importance of somatic change with these two 'case studies'; and explore the possibility that old age and death itself in many single-celled creatures, and in other multicellular organisms, may reflect the genetic decay of the soma.

7.2 Somatic mutations

Occasionally one sees people with one brown and one blue eye. There has been a mutation in one eye as it develops, changing its colour. Somatic mutations were once seen as little more than curiosity. Now, though, it seems that such events are central to the economy of the body in health and in sickness. The molecular and structural changes involved look remarkably similar to those that take place in the germ line.

In fact, *most* mutations occur in somatic cells. With an error rate of 10^{-10} in DNA replication and 10^{15} cell divisions during the lifetime of an adult, many thousands of new mutations must occur somewhere in every genome. Depending on the nature of the mutation, its location, and the tissue involved, somatic mutation may lead to individual differences in phenotype or even to cancer.

In Chapter 4, you saw that the frequency of germ-line mutations increases with parental age. Recall Down's syndrome, for example. The incidence of somatic

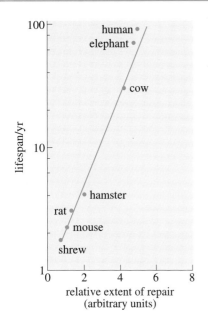

Figure 7.1 DNA repair as a function of maximum life span of mammalian species.

mutations does exactly the same thing; so much so that some elderly people who are not genetically germ line sickle cell carriers even have small amounts of sickle-cell haemoglobin in their blood because of a crucial base change in the gene which normally produces HbA (Chapter 1).

One cause of the increase, in both germ line and somatic cells, is a diminishing ability to repair mutations with age. Cells accumulate damage at rates that depend on the efficiency of DNA repair systems. Their levels are genetically determined; and longer lived species have higher levels of expression. Figure 7.1 shows the correlation between maximum lifespan of various mammals and the ability of cultured cells to repair ultraviolet-induced DNA damage.

Any somatic mutation appears in all the descendants of the mutant cell. Its host becomes a *mosaic* with a mixed cell population; that is, normal cells and mutants coexist in a single individual. Evidence for such somatic mutations can sometimes be seen directly, as in the appearance of mosaic hair colour. For example, a person's hair may be homozygous for brown (*b b*) in most of its cells, but with a white (*B b*) patch. One *b* allele in a cell has mutated to *B* during development.

▷ The size of the white patch varies greatly from person to person. Why should this be?

▶ The later in development the mutation occurs, the smaller the segment of tissue derived from the mutant cell. Thus, some of the fairly common mosaic hair colours in humans are due to late somatic mutations. Think, too, of the patches of mutant cells on the seeds of the maize plants studied by McClintock in her work on transposable elements and mutation (Chapter 4).

Adult somatic cells can accumulate mutations that would be lethal if they occurred early in development. A striking example comes from a case of ornithine transcarboxylase (OTC) deficiency in a boy with an unusually mild combination of the symptoms associated with the disease. His symptoms were limited to a high blood level of ammonia, making him lethargic and damaging his nervous system. The OTC enzyme is essential in the detoxification of ammonia, and its absence is lethal during the early stages of fetal development. The boy was a somatic mosaic for a deletion in the *OTC* gene: some cells had the enzyme, some did not.

▷ Some diseases associated with a breakdown in the mechanism of oxidative phosphorylation share a general tendency towards late onset. What does this suggest?

▶ Somatic mutation damages the respiratory machinery in cells that have undergone many divisions. Studies of mitochondrial DNA in older people (Chapter 6) show that age-related somatic mutation is widespread.

7.2.1   Mechanism of somatic mutation

As in the germ line, genetic changes in somatic cells can be brought about in various ways. Point mutation is only one. Many others exist. For example, loss of the chromosome carrying allele *A* from a cell heterozygous for *A a*, will leave the cell hemizygous for allele *a*.

▷ How can loss of a chromosome arise? (*Hint*: think back to the origin of Down's syndrome.)

▶ Loss of a chromosome in one progeny cell (and the gain of a chromosome in the other), as in Down's syndrome.

Non-disjunction in an early post-zygotic mitotic division is a common cause of mosaicism. Individuals who are mosaic for a particular trisomy, such as mosaic Down's syndrome (having some normal cells and others with trisomy of chromosome 21), are less severely affected than are non-mosaic trisomic indivduals. Some chromosomal mutations are tolerated in somatic cells but not in the germ line. For example, trisomy for chromosome 8, although lethal in the germ line, does occur in fibroblast cells (but not in leucocytes).

Genetic mutation in somatic lineages may also arise from crossing over between homologous chromosomes at mitosis. This process (as in meiosis in the germ-line cells), occurs between two of the strands at the four-strand stage (Figure 7.2). In a human, heterozygous for brown eyes $B\,b$, crossing over in a developing eye cell can lead to the production of $B\,B$ and $b\,b$ progeny cells. Thus, in the example in Figure 7.2, there would be a blue ($b\,b$) spot in an otherwise brown eye (containing, after somatic mutation, both $B\,B$ and $B\,b$ genotypes).

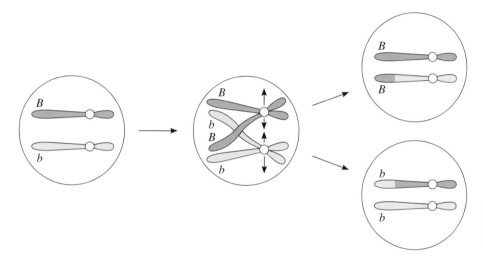

Figure 7.2 Somatic crossing over in a cell of the genotype $B\,b$ and the genotypes of the progeny cells.

7.2.2 Genetic change and the immune system: somatic evolution

This brief overview of the mechanisms of somatic mutation suggests that there is as great a variety of mutational changes as is found in the germ line. In one set of genes in somatic cells, though, change takes place at a rate and with a range of mechanisms seen nowhere in the germ line. As an example of programmed genetic change that has a positive role in the body's economy, we shall look at what is fundamentally a system of **somatic evolution** – the immune system.

The human body has a coordinated system of defence mechanisms which has evolved in response to the threat of infection from pathogens. This network of cells and molecules – the immune system – is hugely complex (as you already know from Book 3). It needs to be: pathogens come in an enormous diversity of shapes and sizes. Their bodies are composed of a wide range of chemicals some of which are specific to them and some shared with the host.

The immune system must hence distinguish between the body's own molecules and those of an invader. It faces another problem, too. The fast generation time of micro-organisms means that their genotype can change rapidly by mutation as they reproduce. The human body must therefore cope both with existing pathogens and with those that may evolve in the future. Even worse, each pathogen has its own mechanism to protect itself from the host's response. The immune system, has not only to be enormously versatile to identify a vast range of molecules, but has to be very precise to identify the subtle distinctions between self and non-self. How does it manage? Somatic change is at the centre of these remarkable and in some ways contradictory functions.

First, the host produces its own cell-surface antigens which define its 'self', and which are involved in interactions between white cells (leucocytes). These antigens identify foreign cells and hence play an important role in the tissue rejection of transplants. These antigens are coded for by the genes of the major histocompatibility complex (MHC; see Book 1, Figures 3.33 and 3.34). The MHC region covers more than 3.5 million DNA base pairs on chromosome 6 (Figure 7.3). It is highly polymorphic and consists of many genes that can, in turn, exist in scores of different allelic forms. The loci involved are so closely linked that they are usually inherited as a single block, a haplotype (Chapter 1). Certain haplotypes are frequent and others rare. Since everyone inherits two haplotypes (one from each parent) no two people have the same set of MHC antigens: everyone (apart from monozygotic twins) has a unique cluster of cell-surface antigens.

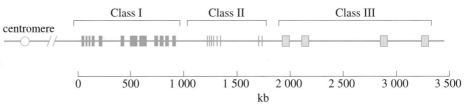

Figure 7.3 A schematic drawing of the major histocompatibility genes on chromosome 6.

The immune response involves defensive cells and proteins (the antibodies) circulating in the blood. It identifies pathogens by the shape of their surface epitopes (Book 3, Chapter 4; Figure 7.4) by means of a huge population of receptors, a few of which make a 'lock and key' fit with the epitopes. Millions of different receptors are required to achieve the immune system's extraordinary sensitivity. As their production is under genetic control this raises an obvious problem. Given that there are only some tens of thousands of coding genes in the genome, the majority of which lie outside the chromosomal segment responsible for the immune system, how can there be specific genes for each of the innumerable receptors produced? The idea of *programmed genetic change* of certain genes – of somatic change – came to the fore in immunology only a few decades ago.

The major histocompatibility complex shows a striking similarity in organization and in DNA sequence to the other genes in the immune system. All belong to a gene family called the immunoglobulin superfamily (Book 1, Chapter 3). Its members are related, and may have descended from a common ancestral gene which duplicated and diverged long ago – before the origin of the vertebrates – to give distinct products. Its best-known members are the antibodies produced by the B lymphocytes and the antigen receptors on the surface of of the T lymphocytes. These specialized leucocytes mature in the **b**one marrow and **t**hymus, respectively.

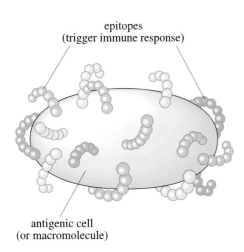

epitopes
(trigger immune response)

antigenic cell
(or macromolecule)

Figure 7.4 An antigen is a large molecule or cell with one or more unique areas, called *epitopes*, in its structure.

Recall from Books 1 and 3 that B cells synthesize and secrete antibodies which respond to invading toxins and to antigens on the outer surfaces of pathogens.

T cells, with complex receptors on their membranes, respond to antigens on pathogens that replicate inside the host's cells. In turn, T and B cells interact with the network of MHC molecules and other cells in the immune system to produce a sensitive surveillance and protection system which – usually – keeps the outside world at bay.

One lymphocyte looks much like another. In reality, though, each is unique, secreting thousands of copies of a specific antibody or carrying thousands of copies of a specific receptor protein on its surface. How do cells produce a myriad of distinct proteins if they all derive from stem cells with the same genotype? It seems scarcely possible that each organism has DNA coding for antibodies to all the possible epitopes on all the 'foreign' cells and macromolecules that might get into the body.

Here we concentrate on the production of antibodies by the immunoglobulin genes in B cells. The discovery of how the human immune system makes 10^8 different antibodies from a limited number of genes emphasizes the importance of somatic cell genetics – that is, genetic change in somatic cells.

As described in Book 1, Chapter 3, the basic structure of an antibody consists of two 'heavy' and two 'light' protein chains. Both have a constant part (C), which is shared by many lymphocytes (Figure 7.5) and a variable part (V), which shows enormous diversity and is specific to each lymphocyte (Figure 7.5). The V region forms the antigen-binding site, which binds to just one specific epitope.

For each of the four protein chains, the V and C regions are joined together to give a single polypeptide. At what stages of protein synthesis does this take place: prior to transcription, at transcription, at translation or post-translation? A crucial clue emerged when single mRNA molecules (isolated from a lymphoid tumour which was antibody-producing) were found to contain sequences for *both* C and V regions.

▷ What can you conclude from this?

▶ The V and C regions of the polypeptide cannot be translated from different mRNA transcripts and then fused into a single protein chain. Instead, the process must take place either at an early stage in transcription, or at the level of DNA prior to transcription.

Are the V and C regions encoded by one gene or separate genes? This problem was resolved by using single mRNA molecules, containing both *V* and *C* regions, as a probe to identify the relevant DNA fragments in the genome. These fragments differed according to whether the DNA was extracted from antibody-producing cells or from those not involved in making the proteins. The labelled RNA hybridized to *two* fragments of DNA extracted from non-producing cells. This suggested that, in non-producing cells, genes specifying V and C regions are separated in the genome. In antibody-producing cells, however, the mRNA hybridizes to only *one* DNA fragment. Moreover, in the undifferentiated stem (parent) cells of the B lymphocytes, the distribution of the antibody DNA sequences is the same as in non-producing cells.

▷ What can you conclude from these experiments?

▶ That *V* and *C* genes are separate in the stem cell but are joined together in B cells as they mature, so that the two genes are transcribed as a single unit.

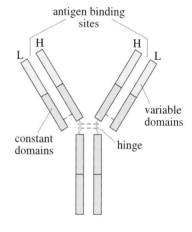

Figure 7.5 Basic structure of an immunoglobulin molecule (H = heavy chain, L = light chain).

Antibodies are encoded by gene clusters, each having a similar structure. The top line of Figure 7.6 shows the structure of a light chain immunoglobulin gene cluster in stem cell DNA. Each spans millions of base pairs and is divided into a number of sections, the variable (V), joining (J) and constant (C) segments.

In stem cells, the V segment (which consists of at least 50 genes) lies many kilobases from the J segment which is made up of a much smaller number of genes. As B lymphocytes mature the gene cluster undergoes somatic rearrangement (recombination): a V region gene and a J region gene join together. The intervening genes and non-coding regions are lost (Figure 7.6a). This produces a V-J-C DNA sequence, which is transcribed (Figure 7.6b). As in other genes, introns in the transcriptional unit between the J and C segments are removed by RNA splicing (Figure 7.6c) to generate a mature mRNA for translation (Figure 7.6d).

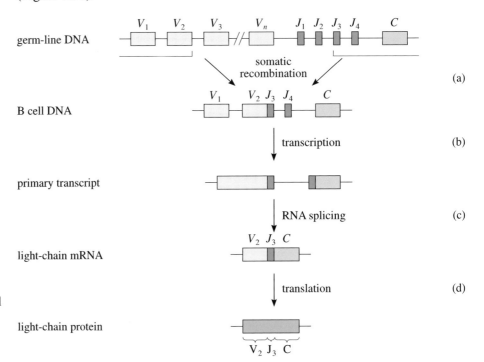

Figure 7.6 Structure of a light chain immunoglobulin gene cluster in stem cell DNA, somatic recombination (a) and other processes (b–d) involved in antibody formation.

The variation present in each of the V-J-C sequences means that this mechanism has the enormous potential to generate antibody diversity. The heavy chain immunoglobulin gene, too, has a similar structure to that of the light chain (except that it has a fourth region of D – diversity – genes, and an even greater number of V and J genes). A great number of combinations are possible. Any V can join with any D, and any D to any J. Somatic rearrangement of this kind is a pre-transcriptional control mechanism. The process, which involves DNA splicing, is an effective way to produce the maximum amount of variation from the minimum number of genes.

Slight variations also occur in exactly where in the nucleotide sequence the cutting and splicing takes place at the 'margins' of the selected genes; this produces further variation in the polypeptide sequence of the product – the variable region of a light or heavy chain. Somatic mutation within the V region genes also increases antibody diversity. Yet more variety in the system results from the assembly of protein chains, two heavy and two light, into a complete antibody. The combination of all these mechanisms yields a large number of possible permutations – more than 10^8 different mature antibodies per genome.

Each B lymphocyte produces, because of this hierarchy of mechanisms, its own unique antibody. Somewhat similar mechanisms occur in T lymphocytes, enabling them to produce a huge variety of antigen receptors. Furthermore, each mature lymphocyte has its own specific DNA sequences with a genotype different from that of its 'parent' stem cell and from all other lymphocytes. Somatic change (some of which involves rearrangements of the genetic material) has generated an enormous genetic diversity within the body of a single individual. This discovery cast a shadow on what had seemed a core precept of genetics: that *all* somatic cells of an individual have the same genotype.

▷ Identical twins have identical genotypes. Do they have an identical repertoire of antibodies?

▶ No: each lymphocyte in each twin has its unique antibody repertoire because somatic rearrangement leads to random combinations of *V-J-C* genes.

Pathogens have evolved their own strategies to counteract the host's immune response. Many exist. Mutation is one strategy. In HIV, the virus that causes AIDS, the genes that encode the surface antigens are unstable. HIV changes its antigens so frequently that even within a single infected individual, new strains of the virus are generated as the infection proceeds (Chapter 5). DNA rearrangement also enables certain other pathogens to avoid detection by the immune system. It takes place, for example, in genes that encode the surface pili of *Neisseria gonorrhoea* (the bacterium that gives rise to gonorrhoea), and the surface glycoproteins in trypanosomes (the parasites that cause sleeping sickness). Sexually reproducing pathogens have an added advantage in that they can produce new combinations of antigens by genetic recombination. The need for a somatic system – the immune system – to combat a sexual system helps to explain why somatic change has reached such extraordinarily high levels in this one section of the genome. Some biologists even believe that sex itself may have evolved as a result of the evolutionary pressure between parasites and their hosts.

7.3   Cancer – somatic mutation and cellular disease

Cancer is essentially a genetic disease of cells. A key difference between cancer and other genetic diseases is that it is, for the most part, caused by *somatic* mutations. Most other genetic diseases (except some of those involving mitochondrial genes) are caused by germ-line mutations. Cancer hence gives a unique insight into the importance of somatic mutation. It is not a single illness, but a wide variety of diseases which can affect any tissue of the body. The somatic changes may involve chromosome number, chromosome structure or point mutations. Some chromosome changes and gene mutations give the cell a growth advantage over unmutated cells and give rise to malignancies. As you will see, the mutations that lead to cancer affect genes involved in cell division and in the cellular signalling network.

The main proof that cancer involves genetic changes is that cancers are *clonal*; they contain the same genetic change that can be traced back to a single cell in which the first step towards cancer took place. For example, there is a consistent chromosome abnormality, the presence of a small chromosome 22 (Figure 7.7) in people with chronic myelogenous leukaemia (CML).

This leukaemia is a malignancy that arises from the precursor of myeloid cells (one of the types of circulating white cells of the immune system). The abnormal chromosome is called the Philadelphia chromosome, after the city in which it was discovered. It is never found in normal cells and is stably inherited in all progeny cells. This suggests that the genetic change is indeed responsible for the **transformation** of normal to cancerous cells. Since it is unlikely that this specific abnormality arises more than once in the development of leukaemia in an individual, its presence is evidence that all the cells in this tumour are genetically identical – they are a *clone*.

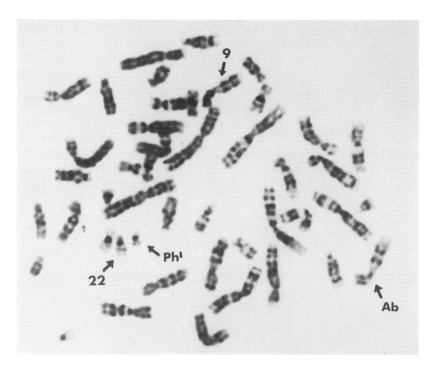

Figure 7.7 Metaphase spread of chromosomes of white cells from a person with CML. Translocation of long arm material from chromosome 22 to chromosome 9 results in a small chromosome 22 known as the Philadelphia chromosome (Ph[1]), and an abnormal chromosome (Ab). Compare this metaphase spread with the normal karyotype in Figure 2.3, Chapter 2.

The clonal origin of cancer does not imply that a tumour originates from a *single* transforming event. If this were so, tumour incidence would increase in a linear fashion with age. Figure 7.8 shows how the incidence of cancer increases with age. This non-linear increase in incidence suggests that the development of cancer involves a progressive *series* of changes at the genetic level.

It appears that as many as six or seven mutations may have to accumulate before a cell becomes cancerous. But as age increases, mutations do accumulate in cells – explaining why cancer is mainly a disease of older people. So a second key difference between cancer and other genetic diseases is that each individual cancer involves, not one, but a series of changes at the genetic level. This 'multi-hit' model suggests that cancers grow by a process of clonal evolution driven by mutation. Many of the mutations may be unimportant. Others may be threatening, particularly those in genes associated with the control of cell proliferation and the cellular signalling network. The challenge is to identify, in any particular cancer, the specific mutations involved.

7.3.1 Genetic changes and cancer

Every human cell contains genes that have the potential to cause cancer. Which ones actually do the damage? Three types of genes are involved. The proliferation of *normal* cells is regulated by growth-promoting **proto-oncogenes** counterbalanced by the growth-constraining **tumour-suppressor genes**. Also important are the genes that promote genetic stability, such as those involved in

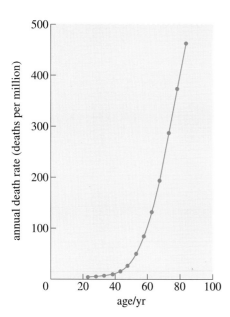

Figure 7.8 Annual rate of death from colon cancer in the United States.

DNA repair (Chapter 4). Studying these three types of genes provides an insight into normal molecular mechanisms of cellular growth as well as how cancer develops when such mechanisms go wrong.

7.3.2 Oncogenes

Oncogenes are derived by mutation from normal cellular growth-promoting genes, the proto-oncogenes. Proto-oncogenes are highly conserved through evolution from yeast to humans. There may be up to a 100 proto-oncogenes and, therefore, potentially that many oncogenes. To understand them demands the investigation of chemical carcinogens, viruses and genes, all of which are involved in the production of cancer. Proto-oncogenes can mutate in a variety of ways.

Chromosome translocations

The Philadelphia chromosome, discussed above, arose as a reciprocal translocation (Chapter 4) between chromosomes 9 and 22 (Figure 7.9).

Molecular analysis reveals a gene, the *abl* proto-oncogene, on chromosome 9 in the region of the break-point, which is transferred to the Philadelphia chromosome itself. The *abl* gene codes for a tyrosine kinase, an enzyme which phosphorylates tyrosine residues in other proteins. The break-point of chromosome 22 is within a gene called *bcr* (for **b**reak-point **c**luster **r**egion). The function of the *bcr* gene is unknown (although it is transcribed in normal cells). The organization of the *abl* and *bcr* genes before and after translocation is shown in Figure 7.9.

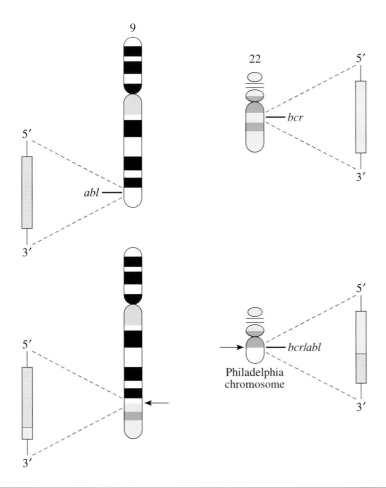

Figure 7.9 The origin of the Philadelphia chromosome, and organization of the genes at the chromosome break-points before and after translocation.

199

▷ Concentrating on the structure of the chromosomes, from what genetic material is the Philadelphia chromosome composed?

▶ The upper end, including the centromere, is from chromosome 22 and its bottom tip is a piece from the tip of chromosome 9.

Subject to the limitations of cytogenetic observation, translocations in different cases of CML always involve the same positions on chromosomes 9 and 22. Which genes are near these break-points and what role might they play in the development of leukaemia?

▷ What happens to the 3′ end of *bcr* after translocation?

▶ It is transferred to the tip of chromosome 9.

▷ What happens to the sequences 3′ of the break-point of the *abl* oncogene after translocation?

▶ They become joined to the part of the *bcr* gene that remains on chromosome 22.

After translocation, the Philadelphia chromosome contains a new *fusion gene*, composed of part of the *bcr* gene and the *abl* gene, whose organization is: centromere-5′-*bcr*-*abl*-3′-telomere (Figure 7.9). The crucial question for cancer development is whether the new gene is transcribed and hence has a function.

The fusion gene is a template for a 'fusion RNA' which is crucial for the development of CML as it allows the gene to be translated to produce a 'fusion protein' consisting of coding sequences derived from two different genes. (Recall the fusion proteins produced artificially in the synthesis of recombinant insulin – Chapter 3.) The fusion protein, which is longer than the normal *abl* protein, is now under the control of the *bcr* promotor and has enhanced tyrosine kinase activity. Further genetic changes are required to transform the clone into a leukaemic phenotype – another example of the need for multiple hits in cancer evolution.

Not all tumour-promoting oncogenes generate a new gene product – albeit with the same function, in this particular case. Changes in proto-oncogene activity are brought about in other ways. In the case of Burkitt's lymphoma (a fast growing malignancy of the B cells of the immune system) a chromosome translocation is again involved, but this time between chromosomes 8 and 14. As you learnt in Section 7.2.2, a major function of B cells is to produce antibodies (immunoglobulins), a group of proteins that bind to 'non-self', or foreign, cells and molecules (antigens) (see also Book 1, Chapter 3, and Book 3, Chapter 4). The genes that encode the production of antibodies, the immunoglobulin genes, are expressed at high levels in B cells. The break in chromosome 14 is in one of the immunoglobulin genes. It presumably happens during somatic rearrangement of this gene. The *myc* proto-oncogene on chromosome 8 (close to the break-point) codes for a nuclear protein normally expressed in small quantities in these cells.

The translocation moves *myc* from chromosome 8 to the immunoglobulin gene (*Ig*) at the break-point of chromosome 14 (Figure 7.10). This destroys normal gene regulation of the *myc* oncogene so that it is constantly expressed at high levels. Its new 'constitutive' status arises from its linkage to sequences that normally enhance immunoglobulin production. The *myc* gene is now expressed in the same way that immunoglobulin genes are expressed in a normal B cell.

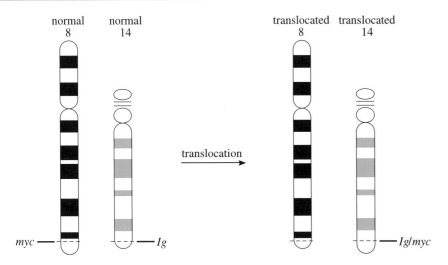

Figure 7.10 Translocation of *myc* from chromosome 8 to chromosome 14 in Burkitt's lymphoma.

Chronic myelogenous leukaemia and Burkitt's lymphoma show how chromosomal abnormalities may be features of malignant cells. These two types of cancer also show how **tumorigenesis** (the production of tumours) often involves translocations which lead either to the activation of proto-oncogenes, or to the creation of tumour-specific fusion proteins. In the case of leukaemia or lymphoma, DNA breaks may be associated with the DNA rearrangements that naturally occur in lymphocyte precursor cells (Section 7.2.2), as in the case of Burkitt's lymphoma. Different types of genes and different mechanisms of change may operate at different stages of tumour development. Only some are visible at the cytogenetic level.

Retroviruses

Viruses attack cells. Sometimes they kill, but some viruses can become integrated into the host's DNA. They are sometimes involved in the disruption of gene function during germ-line mutagenesis. Such viruses, the retroviruses (Chapter 2), are important in cancer development.

Their danger lies in their replicative cycle (Figure 7.11). The genetic material found inside the protein coat is RNA rather than DNA. Nevertheless retroviruses grow and replicate inside cells whose genetic material is DNA.

▷ How is information coded as RNA understood by cellular machinery that reads only DNA?

▶ The incoming viral RNA becomes attached to molecules of the enzyme reverse transcriptase. This enzyme has the remarkable property of 'transcribing' the viral *RNA into DNA* which can then code in the normal manner for viral mRNA.

This reverse flow of genetic information is crucial to the life cycle of the virus. The DNA copy of its RNA message is integrated at random into the cell's chromosomes. It is not detected as 'foreign' by the cell's DNA repair mechanisms and is transcribed by the cell's own RNA polymerase. The resulting RNA does two things: it acts as the genome for progeny virus and as a messenger RNA for the translation into viral proteins. The viral genomic RNA is packaged inside protein coats. Progeny virus bud through the membrane of the infected cell, ready to attack new cells (Figure 7.11). Viral infection causes little apparent damage to the cells themselves.

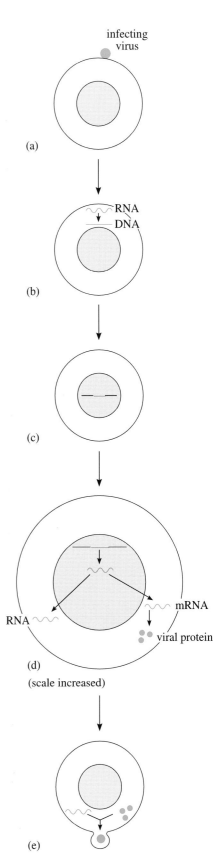

Figure 7.11 The normal replicative cycle of a retrovirus.

Retroviruses can contribute directly to the development of human tumours. This is because of another remarkable property: their ability to capture a piece of their host's genome. Most retroviruses have but three basic genes, sufficient for the retrovirus to reproduce itself inside the host cell. However, the *acute* type of cancer-causing viruses have an extra one. In the case of Rous sarcoma virus (a tumour virus from chickens), for example, the extra gene is called *src* (derived from '*src*oma' – the name given to a particular class of tumours). The presence of *src* leads to uncontrolled cell division and to cancer. Remarkably, sequences homologous to those of the *src* gene are present in humans. During its evolutionary history the virus must have, after integrating into the chromosome of the cell it had invaded, picked up an adjacent human gene – an oncogene – by recombination. When it left its chromosomal position to generate new virus particles (see Figure 7.12) it took the captured gene with it. On infecting a new host cell, the captured human gene is integrated into the chromosome with the 'viral' DNA. There it is transcribed (Figure 7.12).

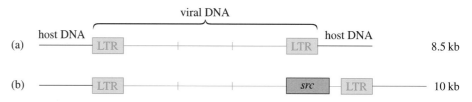

Figure 7.12 Organization of the Rous sarcoma virus genome (b) with the integrated *src* gene, and compared with a non-transforming retrovirus (a). The viral DNA is contained within LTRs (long terminal repeats) which direct the efficient integration of viral DNA into the genome. The length of viral DNA in (a) is 8.5 kb compared with that of (b) which is 10 kb.

A number of oncogenes are picked up by viruses and altered so that they cause cancer. Most proto-oncogenes are named after the virus that carries a DNA sequence homologous to their own (such as *abl* for **Abl**eson murine leukaemia virus). Viral oncogenes are altered versions of their proto-oncogene cellular counterparts, differing from them in regulatory and protein-coding sequences. The oncogenes are often truncated and have accumulated point mutations. These changes in structure and function increase their power to transform normal cells into malignant ones.

Other viruses are less potent in causing damage than are those carrying oncogenes. They can cause cancer but usually take longer to manifest their effects. When integrated into the host's DNA some cause somatic mutation and hence enhance the expression of oncogenes. The oncogene *myc* can be activated by insertion of viral DNA, as shown in Figure 7.13. This process is called **insertional mutagenesis**.

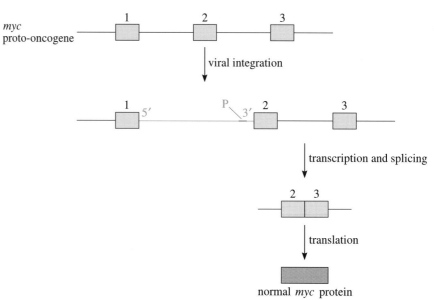

Figure 7.13 Activation of the *myc* oncogene by viral integration. (P = viral promoter.)

The *myc* proto-oncogene consists of three exons. Only exons 2 and 3 are translated. The most frequent site of viral integration into *myc* in lymphomas is between exons 1 and 2 (Figure 7.13). Transcription of *myc* is under the control of the viral promotor at its 3′ end, immediately upstream of *myc* exons 2 and 3 (Figure 7.13). This leads to enhanced expression of the normal *myc* gene. The effect of insertional mutagenesis on gene expression is hence similar in kind to the translocation between chromosomes 8 and 14 in Burkitt's lymphoma.

▷ Recall an event that occurs in the germ line that has parallels to the process of insertional mutagenesis.

▶ Transposition; transposons act independently like viruses and can move around the genome (Chapter 4).

Viruses other than retroviruses may also play a role in the development of cancer. For example, Epstein–Barr virus (EBV) may be involved in Burkitt's lymphoma. The virus readily infects and transforms B cells *in vitro*. It is found in most Burkitt's lymphoma cases in Africa but is missing from the majority of other cases. EBV is hence not necessary for the development of this cancer. In Africa, though, particularly where malaria is rife, Burkitt's lymphoma is much more prevalent than in North America and Europe. The term 'endemic Burkitt's lymphoma' is sometimes used for the African form and 'sporadic Burkitt's lymphoma' for cases found in developed countries.

Since EBV is more or less ubiquitous, why is it not present in all cases of Burkitt's lymphoma? A study of the geography of the disease suggests that other factors may be important. Exposure to malarial parasites is a strong candidate. Children with acute malaria have more EBV than do those without the infection. Perhaps an immune system overtaxed by the fight against the parasite means that those infected more readily succumb to EBV infection. Infections stimulate the production of new B cells. Malarial parasites and EBV may act as 'co-carcinogens'; the production of new B cells is so rapid that the chance of an accident leading to activation of the *myc* oncogene is increased. Burkitt's lymphoma shows the importance of the interaction between environment and genes in the production of cancer, a point to which we shall return.

7.3.3 Tumour-suppressor genes

Certain very rare types of cancer run in families. They are not due to the presence of oncogenes: instead familial cancers are triggered by mutant genes in the germ line. The mutant alleles, therefore, must be present in every cell of the body (whereas oncogenes become activated through somatic mutation that alter genes during an individual's lifetime). The class of genes involved in cancer in these cases are the growth-suppressor or tumour-suppressor genes.

The products of proto-oncogenes promote cell growth, but the products of tumour-suppressor genes normally block it. Only one copy of an oncogene is needed to promote cancer; oncogenes are, in Mendelian terms, dominant. In contrast, tumour-suppressor genes contribute to malignancy only when the normal function of both alleles is lost; tumour-suppressor genes are hence recessive. It is the *loss of function or inactivation* of tumour-suppressor genes, rather than activation (as in oncogenes), that leads to tumour development. For this reason they are sometimes called anti-oncogenes.

A notable example of a tumour-suppressor gene is *Rb,* a gene associated with susceptibility to retinoblastoma, a malignant tumour of the back of the eye. The tumour (which is exceedingly rare, seen in only 1 in 20 000 children) has striking and unusual properties. Retinoblastomas often run in families, as shown in Figure 7.14. The tumour occurs only in young children below the age of 5 years; the retina seems to be a fertile field for tumour growth only up to this age. There are often multiple tumours, and both eyes are usually affected. Treatment usually involves removal of the affected eye(s), although small tumours are removed locally by radiation. Children cured of retinoblastoma grow to adulthood and may have offspring, half of whom, on average, contract the malignancy (Figure 7.14).

▷ Does this pattern of inheritance suggest that *Rb* acts as a Mendelian dominant or recessive allele?

▶ Since half of the progeny are affected it suggests it is dominant.

How can an apparently dominant inheritance pattern of susceptibility to cancer be explained if, as we have just seen, tumour-suppressor genes contribute to malignancy only when both alleles are mutated? The paradox is explained by the fact that children inherit *one Rb* allele from a parent and this allele is present in every cell of the body. One copy of the unmutated gene is sufficient to direct normal growth, so that children born with one normal and one defective (*Rb*) allele develop normally. A second somatic mutation is necessary to knock out the remaining normal allele and produce cancer. When the normal copy of the gene is mutated or lost in a retinal cell during development, deregulated growth begins, leading to a tumour. The cancer's dominance in pedigrees reflects the high probability that the normal allele will undergo somatic mutation in at least one of the original million or so cells that give rise to the retina. Thus heterozygotes for the disorder are likely to be affected, with about 90% of them developing cancer.

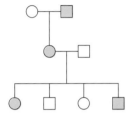

Figure 7.14 A pedigree of familial retinoblastoma.

Tristran da Cunha, a remote island in the South Atlantic, was uninhabited until 1817. Retinoblastoma is very common among its inhabitants. This is not due to the extraordinary environment of the island but to one member of the small founder population who carried the mutation and handed it down to subsequent generations (as described in Chapter 5). Pockets of high frequency of particular cancers hence need not always to be related to exposure to carcinogens.

Retinoblastoma is in fact more complicated than outlined here. There are two types. The same mutant gene is involved in both, and both require two distinct genetic alterations. So far we have considered only the familial type, transmitted from generation to generation because the original mutation arose in the germ line. Around 60% of cases of retinoblastoma, though, are non-heritable; they appear sporadically in an otherwise healthy family (Figure 7.15). In contrast to the heritable type, a single affected eye has a single tumour and the disease appears in later childhood.

▷ Suggest how the sporadic form of retinoblastoma arises.

▶ This type of retinoblastoma arises when *both* alleles of the retinoblastoma gene are inactivated by somatic mutations during retinal development.

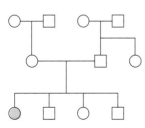

Figure 7.15 A pedigree of the sporadic form of retinoblastoma, for comparison with Figure 7.14.

The need for two independent events at separate *Rb* alleles *within the same cell* explains the fact that there is a single tumour and that it takes a longer time to develop. Retinoblastoma shows particularly clearly that the same mutations can occur in both germ cells and somatic cells. Although for a tumour to develop, in the case of both familial and sporadic retinoblastoma, other mutations are also required.

The retinoblastoma gene was initially mapped by cytological means (Chapter 2). In one patient a deletion of the q (long) arm of chromosome 13 (13q) was found in all cells from a bilateral retinoblastoma (that is, both eyes were affected). In tumours from other individuals with the disease, there were deletions of varying sizes in 13q. Although each deletion is distinct from those seen in tumours from other individuals (as shown in Figure 7.16), they all have a small region in common:13q1.4. (See Chapter 2 for how overlapping deletions are used in mapping.)

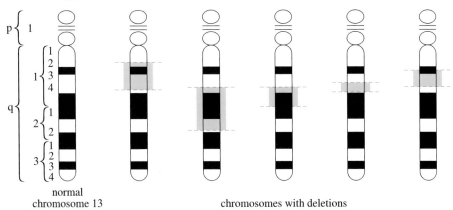

Figure 7.16   Some of the deletions found in patients with retinoblastoma.

▷   What do these observations suggest?

▶   They suggest that loss of a gene on chromosome 13 is involved in the formation of retinoblastoma.

The retinoblastoma gene itself has now been isolated by molecular cloning (Chapter 2). Analysis of DNA polymorphisms in the region close to the *Rb* locus in tumour cells from both heritable and sporadic retinoblastoma patients showed that the genetic change involved is not always a consequence of point mutations or deletions. Although the blood cells of individuals from whom the tumours were taken were heterozygous at the *Rb* locus, the tumours themselves were not heterozygous. (This suggest that the first mutation in these sporadic cases occurred during development, early enough for a number of tissue types to be heterozygous.) The tumour DNA samples contained the *Rb* allele from only one chromosome; it was hemizygous (see Figure 7.17). This revealed a *loss of heterozygosity* for the *Rb* gene. Sometimes this also involved RFLP markers (Chapter 2) closely linked to the *Rb* allele (Figure 7.17). A whole stretch of chromosome 13q in the region of the *Rb* gene had been deleted in the tumour cells.

▷   What could lead to loss of heterozygosity for markers near the *Rb* gene and produce tumour genotypes 3 and 4 in Figure 7.18?

▶   Mitotic recombination (Section 7.2) would lead to homozygosity of the tumour gene (3); chromosome loss due to non-disjunction (Chapters 1 and 4) would lead to hemizygosity of the tumour gene (4).

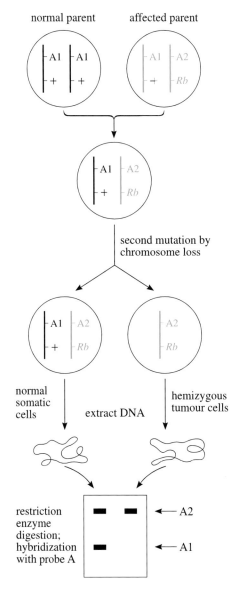

Figure 7.17   Loss of heterozygosity for chromosome 13 RFLPs in retinoblastomas. A RFLP distinguishes alleles 1 and 2 at locus A. An *Rb* mutation linked to marker 2 is inherited and the affected individual is heterozygous at the A locus. + denotes the wild-type allele. (Note: only part of chromosome 13 is illustrated.)

205

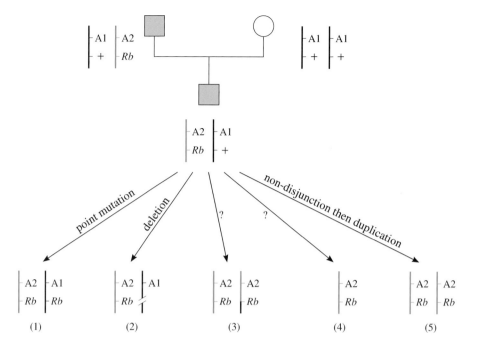

Figure 7.18 Five possible mechanisms for the loss of heterozygosity of the *Rb* allele in the somatic tissue of a heterozygote. The first two are relatively straightforward; they represent a point mutation and a deletion, respectively, of the normal *Rb* allele.

In familial cases that arise due to either mechanism 3 or 4, the retained chromosome 13 is the one inherited from the affected parent – that is, the one with the mutant *Rb* allele. Mechanism 5 in Figure 7.18 is chromosome loss due to non-disjunction (as in 4) followed by duplication of the chromosome carrying the mutant *Rb* allele.

As many as two-thirds of retinoblastoma tumours arise by non-disjunction, with or without duplication, or somatic recombination (that is, mechanisms 3–5), rather than by point mutations or deletions (mechanisms 1 and 2). This is true for both familial and sporadic forms (recall from above that the first mutation in sporadic cases occurred during development, early enough for a number of tissue types to be heterozygous). This is in contrast to cases such as cystic fibrosis (Chapter 8) where affected individuals inherit both mutant alleles through the germ line. Here the majority of the mutations are simple, involving a single nucleotide, while the remainder involve an insertion or deletion of one or two nucleotides, thus changing the reading frame of the mRNA.

Searching for loss of heterozygosity has also helped to identify about 20 other tumour-suppressor genes in the human genome. Some are listed in Table 7.1.

Table 7.1 Some tumour-suppressor genes in human tumours and the tissue most frequently affected.

Disorder	Tumour-suppressor gene	Tissue
retinoblastoma	*Rb*	eye
Wilms tumour	*WT*	kidney
colon carcinoma	*DCC*	colon
neurofibromatosis	*NF*	nervous system
polyposis coli	*MCC*	colon

Rb is a massive gene of 190 kb comprising 27 exons. Substantial portions are missing in many retinoblastoma cells. Some subtle changes in *Rb* gene structure are equally effective in knocking out its function. For example, a single base pair change can bring about complete inactivation.

Just what the *Rb* gene does is not known. Might it be specifically involved in retina development? The fact that the same gene may be mutated or lost in different kinds of tumours suggests not. For example, *Rb* deletions or mutations are also found in about one-third of bladder cancers and almost all small-cell lung carcinomas (as well as other lung cancers) but are absent from other cancers such as colon cancer. Why this should be so is not clear. It is intriguing, since the *Rb* protein is a nuclear phosphoprotein expressed in all cells, that is involved in regulating transcription of other genes. It stops cell growth by binding to a transcription factor and preventing it from doing its normal job of turning on gene expression (Chapter 3). Absence of the *Rb* protein means that the suppression is lifted.

7.3.4 Genetic instability: a third type of cancer gene

Are tumour cells intrinsically more unstable than normal cells? A comparison of the rates at which genetic changes accumulate in tumour and normal cells shows that indeed they are, sometimes by 1 000 times. What is the nature of this genetic instability? A number of factors are involved. They include defects in DNA repair; for example, individuals with the inherited disorder xeroderma pigmentosum (described in Chapter 4) have a repair defect and exhibit high error rates in DNA replication.

Malignant cells are prone to mutate because they are growing so fast, increasing the opportunities for genetic accidents to occur. Their DNA also shows a loss of methyl groups (Chapter 3) which may cause the chromosomes to stick to one another at a time in mitosis when they should be moving apart into progeny cells.

Defects in cell cycle controls (Book 1, Chapter 6) are yet another source of genetic instability. The cell cycle is an orderly sequence of events. It involves genes whose products are essential for progress through the cycle. These include enzymes that duplicate and repair the DNA template and proteins that act as 'checkpoints', monitoring the completion of each event and stopping the cycle if things go wrong.

The *p53* gene (so called because its protein product has an M_r of 53 000) plays a crucial role in this and it acts as a check of DNA synthesis. Mitosis stops if DNA damage is detected, thus preventing a subsequent progression to a cancerous state. Cells exposed to damaging agents such as gamma-radiation, produce large amounts of *p53* protein, which either stops the cell cycle (permitting repair of damaged DNA) or starts the pathway to cell death. This destroys the abnormal clones of cells that could lead to cancer.

Cancer cells with *p53* mutations lose this 'save or destroy' machinery, continue the cell cycle and duplicate the mutated DNA. These mutations, therefore, contribute to genetic instability. Such instability generates multiple genetic alterations leading to cancer.

Cancer phenotypes are complex things and the genotypes involved are equally so. Many important clues as to their causes have been discovered but others remain hidden. Some are coming from an entirely different source: epidemiology.

7.3.5 Epidemiology

Epidemiology is the study of factors that influence the frequency of a condition such as cancer, how it is distributed and how it changes with time. Cancer is one of the biggest killers in industrialized countries, second only to heart disease. Although there has been little change in mortality for the last 40 years, there have been major changes in the type of cancer involved. The current biggest killer is still lung cancer. Deaths in men have started to fall in the UK and the United States as the number of smokers has declined. Deaths from lung cancer continue to rise in women because women took up smoking decades after men did. Deaths from stomach, colon and cervical cancer have declined but deaths from breast cancer, melanoma of the skin and prostate cancer have increased. What can we learn from such statisitics?

One approach is to compare different groups of people and, in particular, to look at migrants between these groups. These comparisons can never prove the cause of cancer but can provide powerful hints. The risk of specific types of cancer differs markedly in different parts of the world. For example, colon and breast cancer, which are among the major types of cancer in the Western industrialized countries, are rare in Japanese in Japan, but not among Japanese living in the United States. A large part of the risk may hence depend on exposure to environmental factors. Associations between the incidence of their cancer and a diet rich in saturated fat and meat are seen in many, but not all, studies. Figure 7.19 shows the association between meat consumption and colon cancer among women in 23 countries.

Figure 7.19 Correlation between intake of meat per person and the incidence of colon cancer among women in 23 countries. The strong positive association suggests that animal fat (saturated) or protein may increase the risk of colon cancer. Abbreviations: JAP, Japan; NIG, Nigeria; YUG, Yugoslavia; COL, Columbia; JAM, Jamaica; CHI, China; NOR, Norway; FIN, Finland; ROM, Romania; SWE, Sweden; PR, Portugal; ISR, Israel; POL, Poland; HUN, Hungary; NET, Netherlands; DDR, East Germany; FDR, West Germany; DEN, Denmark; ICE, Iceland; UK, United Kingdom; CAN, Canada; USA, United States of America; NZ, New Zealand.

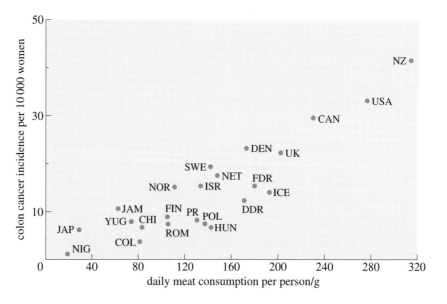

▷ Why are the data in Figure 7.19 not conclusive?

▶ The countries with high meat intake and high levels of colon cancer are affluent Western nations and differ in many ways – only one of which is diet – from the countries with a low-meat diet and low colon cancer rates.

There is, though, much evidence that the risk of colon cancer may be reduced by changing the diet. How can a 'natural' substance in our diet cause cancer? Higher fat intake increases excretion of bile acids and the growth of colonic bacteria capable of converting bile acids to carcinogenic substances, but why meat should have an effect is less clear.

How do we know which substances found in our diet, or environment are safe and which are potentially harmful? Any substance causing genetic damage or increasing the rate of cell division (which in turn increases the chance of a genetic accident), is potentially dangerous. What are the links between environmental hazards and genetic damage?

7.3.6 Mutagens and carcinogens

Percival Pott, an 18th century English surgeon, identified the first environmental cause of cancer. Cancer of the scrotum was common amongst chimney sweeps. Only in the 20th century was it shown that soot contains tars which are carcinogenic.

Many mutagens are, it transpires, carcinogens. To identify them and to assess the risk associated with exposure are matters of public concern. However, such estimates are not easy. There is a vast number of potential mutagens. Which ones are hazards to health? The answers are sometimes surprising.

Which of the following, do you think, makes the most significant contribution to causing cancer: air pollution, discharges from a nuclear reprocessing plant or natural substances found in the diet? The answer is, somewhat surprisingly, natural substances found in the diet.

The Ames test, developed in the 1970s by Bruce Ames, is a quick way to screen for potential mutagens in bacteria. But bacteria are evolutionarily a long way removed from humans. Hence, mice have been used as test systems for human carcinogens. Potency is measured by the daily dose rate required to halve the number of tumour-free animals over a lifetime. Extrapolating from rodent cancer tests done at high doses to humans exposed to low doses is difficult. Nevertheless, tests can indicate that some chemicals are more dangerous than others. They put the hazards of synthetic carcinogens into perspective.

For example, moulds synthesize a wide variety of toxins, of which aflatoxins (Chapter 4) are the most potent of all known rodent carcinogens. A high intake of aflatoxins occurs in those areas of the tropics where peanuts are badly stored and this is associated with high rates of liver cancer.

Rotting food is an obvious source of carcinogens. Some sources, though, are much less predictable. For example, cooking generates a variety of mutagens, all of which are potent carcinogens in rodents. The intake of carcinogens from eating burned or browned food each day is several hundred times more than that inhaled from severe air pollution.

Another way of testing for somatic mutations is to look at the chromosomes of leucocytes in culture. High-risk groups of humans can be identified directly. Some mutations caused by mutagens are indeed involved in the transforming process that results in cancers. Individuals exposed to chemical mutagens at work, such as pesticides and solvents, have more chromosome damage than other workers.

Ionizing radiation from X-rays and atomic explosions (Chapter 4) also increase the amount of chromosome damage such as the number of chromosome breaks.

▷ Would you expect survivors of Hiroshoma and Nagasaki bombings to have an increased incidence of cancer?

▶ Yes. Any agent that increases chromosome damage might be expected to lead to an increased incidence of cancer.

As discussed in Chapter 4, surviving victims of the atomic bombs showed an increased frequency of a variety of malignancies including leukaemia, breast cancer and cancer of the lung and digestive system, compared with the population who had not been exposed. There is a long latency period – from 5 years for leukaemia to up to 40 years for tumours such as those of colon and skin. Why the time lag? One explanation is that the radiation induced a mutation in a cell, but additional mutations in the progeny of this cell were required for a cancer to form. This is yet another example of the need for multiple hits in cancer.

The increased incidence of cancer in survivors of the atomic bombings contrasts with the lack of detectable damage in their germ cells (Section 4.7, Chapter 4). There is no evidence for an increase in inherited mutations, and no increase in cancer has been found in offspring of the bomb survivors.

There is much concern about radiation hazards from nuclear plants. How do the risks compare with those from natural sources? Everyone inhales about 2 000 litres of air each day. Even modest air pollution can result in a person inhaling an appreciable dose of chemicals. Indoor air pollution is often worse than that outdoors. Cigarette smoke is one pollutant but, in some places, the most important indoor air pollutant is radon gas. Radon is a natural radioactive gas that is present in the soil, gets trapped in houses, and gives rise to radioactive decay products. Indoor radon accounts for nearly half the population exposure to ionizing radiation in the UK. The two most important sources are radioactive rocks (such as granite) and cosmic (very short-wave) radiation from space. The dose received by an individual depends on how and where they live. Air travel increases exposure to cosmic radiation; for example, someone who makes five return flights a year to the west coast of North America receives the same radiation dose as received by those exposed to discharges from Sellafield. This dose is itself only about a third of that received by people in Cornwall and Devon where the local granite means that radiation doses are high.

Chemical and physical carcinogens leave traces of their activities on DNA because of the patterns of base changes they induce. Studies of individual mutations at the *p53* locus show how different mutagens act upon different tissues of the body. Mutations in *p53* are the single most common genetic alteration, occurring in 60% of all human tumours. Figure 7.20 contrasts the frequency of the two types of base substitution mutations – transitions and transversions (Chapter 5) – in the *p53* gene from cancers of different tissues. Different mutagens initiate different types of mutations; benzopyrenes found in cigarette smoke and petrol fumes preferentially produce transversions, for example.

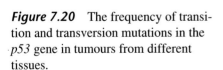

Figure 7.20 The frequency of transition and transversion mutations in the *p53* gene in tumours from different tissues.

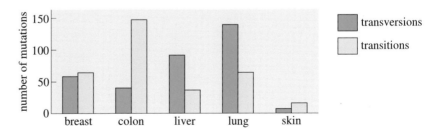

These differences in the ratio of transition to transversion mutations suggest that different mutagens act upon different tissues. For example, lung cancers show more transversion mutations than do colon or breast cancers. This ties in with the action of mutagens in cigarette smoke. Most carcinomas of the skin result from mutations caused by ultraviolet light and are transition mutations. The common liver cancers in southern Africa are, most commonly, associated with a transversion mutation. They arise from the presence of moulds that produce aflatoxin in badly stored peanuts (as discussed above). Where peanut crops are inspected and moulds eliminated, *p53* mutations are much less common.

7.3.7   Gene and environment: a cancer case study

Cancer is caused by the interplay of many environmental and genetic factors. This interplay is illustrated particularly clearly by cigarette smoking and lung cancer.

Most lung cancer deaths result from cigarette smoking, but not everyone who smokes develops lung cancer (or an associated cancer in the mouth or bladder). One reason is that genes influence individual susceptibility. US blacks and New Zealand Polynesians have high lung cancer rates, but low bladder cancer rates. This is because they have genes that give some protection from one carcinogen in cigarette smoke, the aromatic amines that cause bladder cancer. Aromatic amines are metabolized by the enzyme acetyltransferase (and converted into relatively harmless amides). Depending on which alleles they possess, people are either slow or fast acetylators. Fast acetylators quickly convert the aromatic amines into harmless products and are at a decreased risk of bladder cancer.

This is a difference between groups but are there equivalent genetic differences within a group? A class of drug-metabolizing enzymes, encoded by the cytochrome P450 genes (of which there are dozens, perhaps even hundreds), is important in the body's defence against, not just synthetic chemicals, but also those in food and in smoke. They convert them into forms which can easily be excreted.

Many of the cytochrome P450 genes are polymorphic. As a result, there are large variations in drug metabolism within and between human populations. One genetic polymorphism associated with susceptibility to lung cancer is that of the cytochrome P450 enzyme, aryl hydrocarbon hydroxylase (AHH). This is involved in the metabolism of polycyclic hydrocarbons found in cigarette smoke. It converts hydrocarbons into a carcinogenic form. Cigarette smoke actually *induces* the production of the enzyme. Some people carry a 'high-inducible' allele which is dominant to the 'low-inducible' allele. People who smoke and have the dominant allele are at an increased risk of lung cancer. Recessive homozygotes are at a lower risk because they are less effective at converting the hydrocarbons to the carcinogen.

Diet also plays a role in lung cancer. There is an inverse relation between vegetable and fruit intake and cancer (including lung cancer). One suggestion is that β-carotene (found in large amounts in carrots) is a protective factor. Although other nutrients might be involved, patients with lung cancer do have lower levels of β-carotene in the blood.

There is also a strong link between asbestos and lung cancer and mesothelioma (a rare type of lung tumour in the cavity surrounding the lung and not the lung tissue itself). The increase in cancer from asbestos peaked during the 1980s as a delayed result of the exposure of workers (especially in shipyards during World War II). The fine-pointed asbestos fibres lodge in tissue causing cell injury followed by cell proliferation. So the same cancer can be caused by more than one environmental agent.

How do the two agents, smoking and asbestos, interact? Table 7.2 shows the death rate from lung cancer in males in England in the 1960s relative to men who neither worked with asbestos nor smoked.

Table 7.2 Death rate from lung cancer; the risk factors are expressed relative to men who neither worked with asbestos nor smoked.

Working with asbestos	Smoker	Death rate from lung cancer
+	-	5 times higher
-	+	11 times higher
+	+	53 times higher

Men who worked with asbestos, smoked or did both had a much increased chance of dying from lung cancer.

▷ What is the combined effect of the two hazards?

▶ The effect of the two hazards combined is not merely added together but is multiplied.

Materials in the environment are not without hazard. There is a high incidence of mesotheliomas in two villages in central Turkey. Very fine fibres (similar to the ones found in asbestos) are present in the rock used for building.

Although cancer can be described as a genetic disease, its development is a result of a series of complex interactions between environmental and genetic factors. The lifestyle of the individual, such as smoker or non-smoker, diet and exposure to ultraviolet light, plays a part. But we need to know more about how carcinogens affect the genetic material, how much and how frequently the human population is exposed to the agent, and what the long-term consequences are for the population.

7.4 Somatic mutation and ageing

Cancer – now one of the major causes of death in Western societies can be regarded as a failure of the ageing process in cells. Cancer cell lineages become immortal, usually because of somatic mutation.

Ageing is, basically, a failure of the cell's machinery after many cell divisions. There are many theories of why organisms are not – at least potentially – immortal. One of these relates to the accumulation of somatic mutations with age.

We saw in the previous chapter that one part of the somatic genome – that of the mitochondria – can accumulate mutations with age, and that this may itself produce some of the diseases (such as certain forms of diabetes) generally associated with old age.

Not only mitochondria show the malign effects of age. Telomeres, the ends of the nuclear chromosomes, also change. These structures carry no genes, but are vital to the chromosome's survival. Broken ends of chromosomes are unstable and either combine with other chromosomes or are degraded. Recall from Book 1, Chapter 5 that the nucleotide sequence of telomeres is much the same in all vertebrates. Each human telomere consists of 250–1 500 repeats of the sequence TTAGGG.

Maintenance of telomere length is dependent upon the enzyme telomerase, which synthesizes telomeric DNA; without it, chromosomes would shorten each time they divided. In fact, in somatic cells, the number of repeats does decline and the telomeres shorten progressively with age. What is more, there is little or no telomerase activity in old cells. In the germ line, on the other hand, telomeres do not become shorter because telomerase activity persists. This keeps them complete and ensures that chromosomes are transmitted undamaged to the next generation. Whether the shortening of telomeres leads to the symptoms of age or whether it is itself a symptom of cellular ageing, is not known.

Cells proliferate only for a limited number of generations (although the 'clock' that regulates this has not been found). Even cells in culture senesce and die. Possibly telomere loss is itself a 'mitotic clock'. Cancer cells escape from the controls that limit the number of divisions and are effectively immortal. Remarkably, tumour cells do have telomerase activity; telomerase activation may hence be involved in the immortalization of cells.

Genetics is providing intriguing insights into the nature of individual human beings – or any other creature. We are, like it or not, each evolving systems on an inevitable pathway of genetic decline which begins at fertilization and ends at death. The most puzzling, and least understood, aspect of genetics, is how one small group of cells – those in the germ line – avoids this decline and is renewed each generation.

Objectives for Chapter 7

After completing this chapter you should be able to:

7.1   Define and use, or recognize definitions and applications of, each of the terms printed in **bold** in the text.

7.2   Briefly outline the effect that somatic mutations may have on the phenotype.

7.3   Using the immune system as an example, explain why all somatic cells of an individual do not share the same genotype.

7.4   Use examples to illustrate each of the different ways in which proto-oncogenes can mutate into oncogenes: chromosome translocations, hijacking by retroviruses, insertional mutagenesis and environmental factors.

7.5   Describe the role of tumour-suppressor genes and, using retinoblastoma as an example, explain (a) the ways in which genetic change is brought about, and (b) the differences in origin between the sporadic and familial forms.

7.6   Describe, using particular examples, how cancer evolution involves a series of genetic changes.

7.7   Describe the genetic and environmental factors that influence the production of cancer, and discuss how the interplay between these two factors explains the distribution of different types of cancer (with reference to particular examples) between groups.

Questions for Chapter 7

Question 7.1   *(Objective 7.2)*

Explain why the size of a mosaic patch of hair varies from person to person.

Question 7.2 *(Objective 7.3)*

Explain how the immune system makes millions of different antibodies from a limited number of genes.

Question 7.3 *(Objective 7.4)*

Give an example of how a chromosomal translocation can change the activity of an oncogene (or proto-oncogene).

Question 7.4 *(Objective 7.4)*

Describe how an oncogene captured by a retrovirus, can contribute to the development of cancer.

Question 7.5 *(Objective 7.5)*

Using retinoblastoma as an example, explain how loss of heterozygosity in DNA from tumour cells can be used to identify tumour-suppressor genes.

Question 7.6 *(Objective 7.6)*

Describe two lines of evidence indicating that cancer involves a series of changes at the genetic level.

Question 7.7 *(Objective 7.7)*

Explain why not everyone who smokes cigarettes develops lung cancer.

Cystic fibrosis

8.1   The disease

Cystic fibrosis is the most common autosomal recessive condition affecting European white-skinned people. In the UK it affects approximately one child in every 2 500 and approximately 5% of the population are carriers. Symptoms include pancreatic enzyme deficiency, chronic pulmonary disease, male sterility and abnormal chloride ion concentrations in sweat. All are consistent with reduced chloride ion transport across epithelial cell membranes causing increased intracellular and decreased extracellular chloride ion concentrations. This imbalance leads in turn to increased water absorption into the cells by osmosis from the exterior and so gives rise to abnormally viscous mucus on the cell surfaces. The sticky mucus may clog alveoli, causing an environment in which bacteria thrive, or block the pancreatic duct, preventing the transport of digestive enzymes to the gut, and may also obstruct the bowel in the new-born. In boys the mucus blocks the vas deferens (the tube through which sperm pass out of the testicle) causing it to degenerate and disappear. However, before the mutant gene was discovered, although abnormalities in chloride transport were known to be involved, no altered or missing ion channels were found and it was considered possible that the chloride ion transport deficiency might be secondary to some other underlying metabolic problem. Treatment of cystic fibrosis involves replacement of the missing pancreatic digestive enzymes by dietary supplementation together with vigorous and painful physiotherapy to dislodge and remove the mucus which would otherwise clog the lungs.

8.2   The 'positional cloning' approach to finding the cystic fibrosis gene

The traditional approach of geneticists, from the beginning of the molecular genetics revolution in the late 1970s, was to identify a gene by virtue of some characteristic which could be associated with its product – usually a protein. For instance, the α- and β-globin genes code for the two polypeptide chains that are the major constituents of haemoglobin (Chapter 1).

▷   Where would you expect these genes to be strongly expressed and how would you use this information to clone the genes?

▶   The α- and β-globin mRNAs should be abundant in red blood cell precursors (although possibly not in the red blood cells themselves since they have no nuclei). cDNA libraries (Chapter 2), made from RNA prepared from bone marrow which is rich in these cells, would contain globin cDNAs at frequencies reflecting the abundance of the globin mRNAs. By screening the library with a probe made from the mRNA, the clones representing the most abundant transcripts can be selected and sequenced. Those which could be translated to the known globin amino acid sequence could be readily identified.

In other cases too, the normal route of investigation in molecular genetics followed a similar path: establishment of a biochemical defect, identification of the enzyme responsible and isolation of the mutant gene. Once the gene was found, human geneticists, following the traditions laid down by *Drosophila* and mice geneticists, would go on to find its position in the genome by linkage analysis or somatic cell genetics (see Chapter 2).

As the science of genetics progressed, many more diseases were studied in which the underlying biochemical defect was, to say the least, obscure. In a disease whose symptoms provide no hint as to the nature of the defective protein, or even as to where it is made, what can be done to identify the mutant gene? The answer was found by turning the conventional logic of genetics upside down. A new approach, initially called 'reverse genetics' was born. The term itself has now almost disappeared; far from being an unusual approach 'reverse genetics' has become the normal way of cloning genes. The term '**positional cloning**' has replaced it.

This method makes few assumptions. The pattern of inheritance of the gene is traced in affected families and, in the same families, the inheritance of previously mapped polymorphic genetic markers spaced throughout the entire genome is also followed. If one of these loci is consistently inherited along with the disease gene, then mutation and marker locus must be relatively close together; a first step to finding the gene itself. The area in which the gene must be is narrowed down by looking for shared inheritance between the disease allele and all available polymorphic markers in the vicinity. Often it is necessary to find new markers until the target interval is small enough to allow **candidate genes** (genes located within the defined interval and likely to be the ones at fault) to be isolated. If a candidate gene differs in structure in those individuals with the disease compared to those without the disease, then the gene has indeed been found.

This approach leads from map position to gene identification and then to the prediction of a gene's product from its DNA sequence. Sometimes it is even possible to suggest how mutations in the protein may cause the symptoms. This logic is the precise opposite to that used earlier in genetics.

The first gene to be cloned solely on the basis of positional information with no other hints available (such as those emerging from an association of specific chromosomal rearrangements with the disease) was the gene responsible for cystic fibrosis (*CF*). The power of the positional cloning method is shown by the astonishing speed with which this gene was mapped and identified. The story of its cloning is told below. While reading it you should bear in mind that the search was not an orderly, logical, careful, scientific progression. Rather, it was a crazy race between competing laboratories. Prestige and funding were at stake. To obtain funds it is necessary to publish results to prove that something has been accomplished with the previous grant. What is published must be carefully judged so as not to reveal too much information to rival research groups. Behind the rather dry scientific story is a far more interesting tale of scientific politics. Reconstructing the actual course of the discovery is almost impossible. In this account, the stages of the search are recounted in a logical manner but, in reality, almost all these experiments were going on simultaneously (along with many fruitless experiments which, though unpublished, helped the hunt to progress).

8.2.1 Genetic linkage

The number of polymorphic loci whose map positions are known is increasing with astonishing rapidity. At the time of writing (1995) there are 10 468 scattered throughout the human genome. Even in 1985 when the first *CF* linkage was investigated there were 346 and it was obviously impossible to test every one of them in families with the *CF* gene. The most logical approach would be to test loci that are regularly spaced throughout the genome.

▷   What is the optimum spacing of the test loci?

▶   This is not easy to deduce. The answer is a trade off between spacing the loci too widely, so that linkage is difficult to prove because the closest markers to the gene are still not very close (and therefore require testing in many large families), and spacing them too narrowly so that many more loci have to be tested (albeit in fewer and smaller families). The human genome is about 3 000 cM in size (see Chapter 2) and if, for example, such loci were 30 cM apart the *CF* gene would be bound to be within 15 cM of one of them and 100 loci would have to be tested to scan the entire genome.

Scientists, being human, are not strictly logical. They followed hints and clues from other sources and tested the loci that were easiest to score. For several years, no linkage was found with any marker. Disappointing though this was, at least it identified segments of the genome where the *CF* gene was not. By the mid 1980s the region of the genome from which *CF* could be excluded grew to 40%. Eventually, in 1985, a linkage was announced. Ironically it was not to a DNA polymorphism but instead to a polymorphism of the electrophoretic mobility of the protein PON (serum paraoxonase). This did not help much because the *PON* gene itself, like *CF*, had neither been cloned nor assigned to a chromosome. However, it was encouraging. Shortly after this, at a meeting of the American Society for Human Genetics, a Canadian geneticist, Lap Chee Tsui, announced the first DNA linkage – to a DNA marker *D7S15* of unknown function (an 'anonymous' marker). There were a lot of urgent long distance telephone calls made from the back of the meeting hall, and within a week several other laboratories announced their own findings of linkage to this small region of the genome.

To find a linkage was only the first – albeit perhaps the most important – step (Figure 8.1, stage 1). Once everyone knew where to look, the successive steps happened relatively quickly. The initial discovery of linkage reduced the region of the genome to be searched from 3 000 cM to about 30 cM. All the markers known to be linked to *D7S15* were quickly tested in the families with the *CF* gene and the target area was reduced (Figure 8.1, stage 2).

As more markers were tested, haplotypes (single chromosome combinations of linked marker alleles, see Chapter 1) could be built up showing the alleles present along all the individual chromosomes in the families. From the haplotypes the positions of recombination events between particular markers could then be pinpointed. Wherever a recombination event occurred between the chromosomes of an individual heterozygous for *CF* the mutant allele could be seen to segregate with the alleles on one side only of the recombination breakpoint. Its position could hence be excluded from the other side (Figure 8.1, stage 3). In this way *CF* was inferred to be **distal** to (on the telomere side of) the gene *MET* and **proximal** to (on the centromere side of) the anonymous DNA marker *D7S8*. *MET* and *D7S8* were approximately 2 cM apart and had previously been located on the long arm of chromosome 7 in band q31 by linkage to markers mapped by *in situ* hybridization to metaphase chromosomes (Chapter 2).

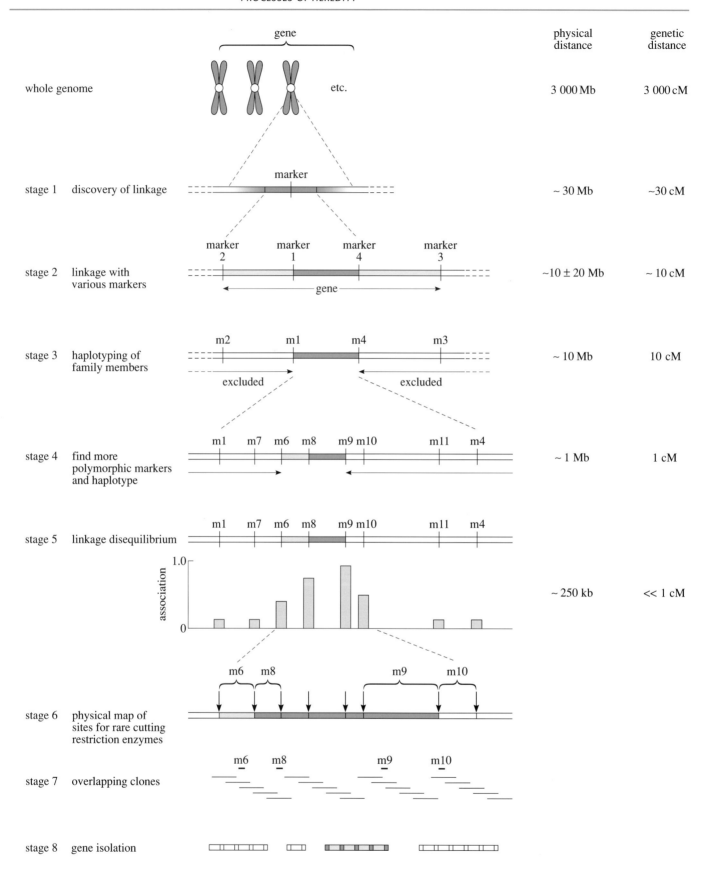

		physical distance	genetic distance
whole genome		3 000 Mb	3 000 cM
stage 1	discovery of linkage	~ 30 Mb	~30 cM
stage 2	linkage with various markers	~10 ± 20 Mb	~ 10 cM
stage 3	haplotyping of family members	~ 10 Mb	10 cM
stage 4	find more polymorphic markers and haplotype	~ 1 Mb	1 cM
stage 5	linkage disequilibrium	~ 250 kb	<< 1 cM
stage 6	physical map of sites for rare cutting restriction enzymes		
stage 7	overlapping clones		
stage 8	gene isolation		

Figure 8.1 The stages involved in the positional cloning of the cystic fibrosis gene. See facing page for details of stages.

Stage 1: The initial discovery of genetic linkage. The area to be searched is reduced from approximately 3 000 Mb to approximately 30 Mb.

Stage 2: Linkage analysis with all the local genetic markers. The search is narrowed to about 10 Mb but possibly with wide limits of error. Multipoint linkage studies will give a probability distribution across the region showing the relative likelihood of the gene being at any particular point.

Stage 3: Haplotyping family members – the search for informative recombination events. The limits of error are narrowed considerably. The closest markers definitely flanking the disease gene are identified.

Stage 4: Discovery of more polymorphic probes from the target region followed by their haplotyping in the crucial recombinants and their families. Luck will play a large part here. The gene may be localized to a region of about 1 Mb. There are nearly always a few polymorphic markers within the target region which are uninformative in the crucial recombinants.

Stage 5: Linkage disequilibrium studies. This enables use to be made of the polymorphisms in the region which were uninformative at stage 4. If the gene has mutated only rarely, there will be a gradient of linkage disequilibrium diminishing on either side of the mutation.

Stage 6: Pulsed-field gel electrophoresis is used to construct a long-range map of the interval from which exact distances may be measured.

Stage 7: Contig assembly. The interval is cloned and candidate genes are sought.

Stage 8: Isolation and identification of the gene.

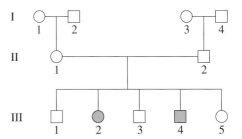

Figure 8.2 The pedigree of an imaginary family in which cystic fibrosis is present in some members.

Table 8.1 The marker loci *MET*, *D7S340*, *D7S122* and *D7S8* have been typed in each family member of the pedigree in Figure 8.2.

I1		I2		I3		I4	
MET	2,2	*MET*	1,1	*MET*	1,2	*MET*	1,1
D7S340	2,2	*D7S340*	1,1	*D7S340*	1,2	*D7S340*	1,2
D7S122	1,2	*D7S122*	1,2	*D7S122*	1,2	*D7S122*	2,3
D7S8	2,3	*D7S8*	1,1	*D7S8*	1,3	*D7S8*	1,4

II1		II2	
MET	1,2	*MET*	1,2
D7S340	1,2	*D7S340*	1,2
D7S122	1,2	*D7S122*	1,3
D7S8	1,2	*D7S8*	3,4

III1		III2		III3		III4		III5	
MET	1,2	*MET*	1,1	*MET*	1,2	*MET*	1,2	*MET*	2,2
D7S340	1,1	*D7S340*	1,2	*D7S340*	2,2	*D7S340*	1,2	*D7S340*	1,2
D7S122	1,2	*D7S122*	2,2	*D7S122*	1,3	*D7S122*	2,3	*D7S122*	1,1
D7S8	1,3	*D7S8*	1,4	*D7S8*	2,4	*D7S8*	1,4	*D7S8*	2,3

▷ Look at Figure 8.2 and the accompanying table. Assuming that genetic recombination will be infrequent in this small interval of the genome, use this information to infer the haplotypes present on each of the chromosomes in the pedigree.

▶ See Figure 8.3

individual	I1	I2	II1	III1	III2	III3	III4	III5	II2	I3	I4
MET	2 ‖ 2	1 ‖ 1	1 ‖ 2	1 ‖ 2	1 ‖ 1	2 ‖ 1	1 ‖ 2	2 ‖ 2	2 ‖ 1	2 ‖ 1	1 ‖ 1
D7S340	2 ‖ 2	1 ‖ 1	1 ‖ 2	1 ‖ 1	1 ‖ 2	2 ‖ 2	1 ‖ 2	2 ‖ 1	1 ‖ 2	1 ‖ 2	1 ‖ 2
D7S122	2 ‖ 1	2 ‖ 1	2 ‖ 1	2 ‖ 1	2 ‖ 3	1 ‖ 3	2 ‖ 3	1 ‖ 1	1 ‖ 3	1 ‖ 2	2 ‖ 3
D7S8	3 ‖ 2	1 ‖ 1	1 ‖ 2	1 ‖ 3	1 ‖ 4	2 ‖ 4	1 ‖ 4	2 ‖ 3	3 ‖ 4	3 ‖ 1	1 ‖ 4

Figure 8.3 The haplotypes present on each of the chromosomes; the affected chromosomes are coloured.

One individual, III4, must be the result of a recombination event between *MET* and *D7S340*. Given that this boy has the disease we can deduce that the *CF* gene must be distal to *MET*.

8.2.2 Linkage disequilibrium

The final step for improving the genetic map depended on population genetics. Once every available family had been tested with every polymorphic probe, the limit of direct genetic mapping had been reached. All the recombination break-points in the families had been defined and the gene location had been narrowed down to the smallest possible interval by examining these recombinations (Figure 8.1, stage 4).

Further information was, however, still hidden in the data. If a mutation is extremely rare then it probably occurred once on one single chromosome in human history. All existing mutant chromosomes would then be descendants of that event. As each copy of the mutant gene has descended from a common ancestor, any recombination events – rare though they might be – would break the original haplotype on which the mutation had occurred. The closer a polymorphic site in the haplotype is to the gene itself, the less likely it is to have recombined away from it. It is hence possible to estimate the relative position of gene and markers by asking if the gene is more frequently associated with one haplotype than with others. Within that haplotype, one can ask whether the gene is more tightly associated with one interval within the haplotype than with others. Of course, if a mutation has occurred several times then it may be associated with several haplotypes, each descended from a different original chromosome 7. The persistence of an association between two alleles like this, whether of disease gene and anonymous marker locus or between two anonymous loci, is known as **linkage disequilibrium**.

▷ Why would haplotypes not provide any useful information if the disease resulted from a frequently occurring mutation?

▶ If the mutation occurs very frequently then it will occur on all types of chromosome and will not be associated with any particular haplotype.

In the case of cystic fibrosis (at least in those patients in which the disease was most severe) the disease did appear to be associated with a particular haplotype. Furthermore, although there was some association with an allele of the *MET* oncogene (Chapter 7), the association was much stronger with alleles of new markers, *D7S340* and *D7S122* (see below), located between *MET* and *D7S8*. The gene was most probably to be found very close to these new loci (Figure 8.1, stage 5).

8.2.3   Narrowing the gap

As the genetic map, based on recombination, was being improved, attention began to be focused on the physical map of the region. The genetic map can only include markers that are polymorphic and between which recombinants are found. The physical map though, can include any DNA locus which has been cloned, polymorphic or not (Figure 8.1, stage 6). The distance scale of a genetic map is measured as the frequency of genetic recombination between markers. At short genetic distances this becomes almost impossible to measure as recombination happens too infrequently. Physical map distances, though, are measured in the numbers of base pairs between markers. These distances become easier to measure the shorter they become. In the search for *CF* there was hence an urgent need to identify more cloned markers from the region to improve both physical and genetic maps.

Various methods were used to find them. One British laboratory obtained cloned probes from the target region of chromosome 7 by using the method of **chromosome-mediated gene transfer** (Figure 8.4) to create somatic cell hybrid lines (Section 2.3.2) each of which contained only short regions of human DNA. One such cell line contained relatively little human DNA except for the region near to the *MET* oncogene (which was, as we have seen, already known to be near the *CF* gene). It was used as a source from which to clone human DNA in the hope that the *CF* gene was within it.

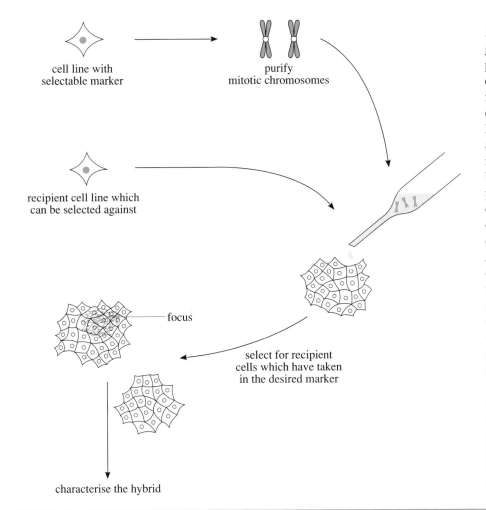

cell line with
selectable marker

purify
mitotic chromosomes

recipient cell line which
can be selected against

focus

select for recipient
cells which have taken
in the desired marker

characterise the hybrid

Figure 8.4   Chromosome mediated gene transfer is a method designed to produce somatic cell hybrids which contain as their only human component a fragment of DNA from the desired region of the genome. Human cells from a cell line with a selectable marker, e.g. activated *MET*, are blocked in mitotic metaphase by the addition of colchicine, broken open and and the chromosomes purified. These are then precipitated with calcium phosphate and dropped onto cells of a recipient rodent cell line which can be selected against, e.g. unactivated *MET*. In these circumstances, some cultured cells possess the ability to take up DNA (or indeed, whole chromosomes). In this case, because the human chromosomes came from a tumour in which the *MET* oncogene had been activated, the mouse cells that had incorporated the appropriate human DNA were converted into rapidly growing tumour cells. The human chromosomes did not survive this adventure intact but tended to fragment. In the resulting hybrid cell lines only short regions of human DNA centred on the *MET* oncogene were found.

A second clever idea was used here in order to bias the markers which were isolated to those coming from genes rather than from inter-genic regions. A type of cloning was used that favoured the isolation of genomic fragments from the CpG islands (which are often associated with the 5′ ends of genes, see Section 2.6.1). This gave several clones derived from the correct region of 7q31 and containing a CpG island at one end. Many were useful for physical mapping and one contained the 5′ end of a gene which, from the mapping data, appeared to be a very strong candidate for the *CF* gene. (Remember the convention – CF refers to the phenotypic condition of the disease, *CF* refers to the gene itself.) This gene was sequenced and was named *IRP* (*int*-related protein) because of its similarity to a known *Drosophila* gene (called *int*). Unfortunately, it turned out not to be the *CF* gene. The proof, negative though it was, came from failing to find mutations in *IRP* in patients with cystic fibrosis, a time-consuming procedure.

At the same time, other research groups were busy isolating clones and mapping in this region. Among many clones from the region were the two polymorphic markers (*D7S122* and *D7S340*) which showed strong linkage disequilibrium with *CF*.

8.2.4 Development of a physical map

Genetic mapping had by now narrowed *CF* down to the small interval between *MET* and *D7S8* (and probably to the vicinity of *D7S122* and *D7S8*) (Figure 8.5). How big is this region in physical terms? The average conversion figure for the whole human genome is a very convenient 1Mb per cM (Section 2.4.2). However, because the rate of recombination is not constant, this figure can vary locally by anything up to ten times. Several rival groups, each using their own marker probes, began constructing restriction maps (Section 2.2.4) of the region using enzymes such as *Not*I. This enzyme and several others like it (whose restriction sites are CG rich and which contain the dinucleotide CpG) digest human genomic DNA at infrequent intervals giving fragment sizes which are up to several million base pairs long. Those most useful for physical mapping are in the 100–500 kb range. This stage is illustrated in Figure 8.1, stage 6. Figure 8.5 is a map which shows the positions of genetic and other markers in the region. The combination of the genetic and physical mapping data defined an interval of 1.5 Mb within which the *CF* gene must lie. The linkage disequilibrium data had suggested that the most likely region in which to find it was distal to the pair of markers *D7S340* and *D7S122* (which turned out to be only 10 kb apart) and proximal to *D7S8*.

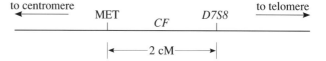

Figure 8.5 A simple map of the gene corresponding to stage 2 of Figure 8.1.

The next step was to obtain as much as possible of this interval in the form of cloned pieces of DNA. At the time, the conventional approach (which had been much used in organisms with smaller genomes such as *Drosophila*) had been to walk (Box 8.1) through the genome. However, given the length of the journey, this would have been a time-consuming process for *CF*. Instead the relatively untried method of **chromosome jumping** (Box 8.2) was used. This generates many small cloned fragments at intervals across the target region. Each of these can serve as the origin of a new bidirectional walk. The more start points that exist in the region, the quicker all the walks can be joined together to make a complete map.

Box 8.1 Chromosome walking

Starting from an appropriate clone a probe is made which, when used to screen a Southern blot (Box 2.2), reveals only a single site of hybridization in the whole genome. The probe is used to screen genomic libraries (Section 2.3.4) to find clones which contain (and hence overlap) it. The most suitable library is usually an array of 10 000–50 000 cosmid or P1 clones derived from the target chromosome. In the late 1980s amplified genomic libraries based on phage λ were still in common use as were amplified whole genome cosmid libraries.

DNA is prepared from the clones hybridizing to the probe, digested with a restriction enzyme such as *Eco* RI and the digest patterns of each clone are then compared. In this way it is possible to align the clones and to identify which one protrudes furthest at each end of the group. Fresh probes can then be isolated from each end and the cycle repeated. If all goes perfectly (an impossible dream) the cycle can be turned once every two weeks. The average advance will be about 20 kb. To close a 2 Mb gap by **walking** from the ends will take about 50 steps from each end, roughly two years of effort. Walking can encounter practical difficulties. Some human DNA sequences, for unknown reasons, do not clone well in *E. coli*. A walk coming up against such a sequence comes to a standstill. The opposite problem is encountered if a walk enters a region which is repeated at more than one site in the genome. In this case, the end probe will identify genomic clones from all other sites and the walk will come to a stop while the bewildered investigator tries to figure out where each clone comes from so that the walk can continue at the correct locus and not accidentally jump to some other chromosome. If the cloned DNA sequence is repeated at more than just a few sites this problem can be insurmountable.

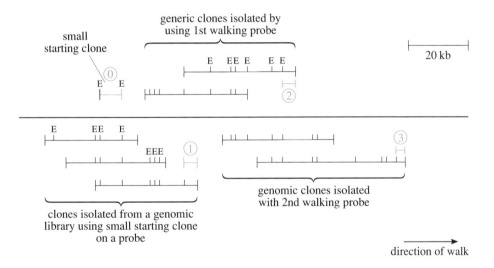

Figure 8.6 A schematic representation of the chromosome walking technique; 0 denotes a small starting clone, 1 the first walking probe, 2 the second walking probe, and 3 the third walking probe, E = enzyme restriction sites for the enzyme *Eco* RI.

Box 8.2 Chromosome jumping

The technique of chromosome jumping was introduced partly to overcome the problems met on walks but also as a way of moving rapidly across a region of the genome leaving a trail of cloned stepping stones behind. Each of the stepping stones can serve to initiate a new bidirectional walk.

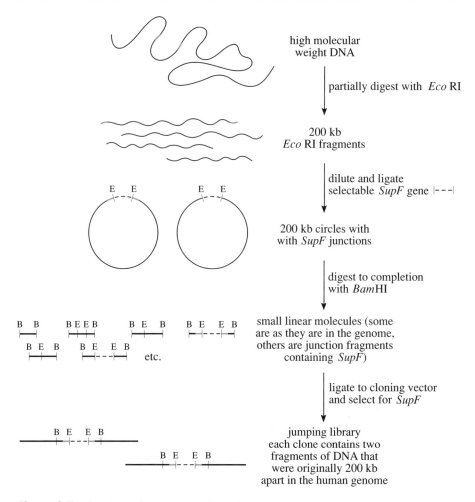

high molecular
weight DNA

partially digest with *Eco* RI

200 kb
Eco RI fragments

dilute and ligate
selectable *SupF* gene |---|

200 kb circles with
with *SupF* junctions

digest to completion
with *Bam*HI

small linear molecules (some
are as they are in the genome,
others are junction fragments
containing *SupF*)

ligate to cloning vector
and select for *SupF*

jumping library
each clone contains two
fragments of DNA that
were originally 200 kb
apart in the human genome

Figure 8.7 A schematic representation of the stages involved in chromosome jumping; the coloured regions denote fragments of DNA.

Starting from high M_r genomic DNA a partial digestion is carried out with an enzyme such as *Eco* RI which cuts DNA frequently. The digestion is carefully controlled to produce an average fragment size of the length which it is desired to jump (usually about 200 kb). The DNA is then diluted and the ends are joined using a ligase enzyme. When the DNA is dilute, the most likely result will be the formation of circular molecules. A high concentration of a very short (~250 bp) DNA molecule containing a gene (*SupF*) which can later be selected is included in the circularization reaction. Many of the circles will incorporate this molecule at the junction. The DNA is then digested with a second frequently cutting enzyme such as *Bam*HI. Some of the resulting *Bam*HI fragments will contain the two ends of the original 200 kb fragment flanking a *SupF* gene. The *Bam*HI

fragments are then cloned in a phage λ vector and used to infect a host strain of *E. coli* in which *SupF* can be selected. The only phage which are able to grow are those that have picked up an insert containing this gene and hence two fragments of genomic DNA which are normally found 200 kb apart in the genome. If the resulting library is screened with a probe from a unique locus, the phage that hybridize to it will contain not only the corresponding locus but also a fragment of DNA from 200 kb away. Because there is some latitude in the sizing of the 200 kb molecules, problems with short unclonable regions do not occur. The jump will almost always find a clonable spot on which to land. Repeated sequences, which make walking such a problem, can create difficulties but more often can simply be jumped over. A number of jumps can be seen in Figure 8.7.

The result of the jumping and walking was a group of overlapping clones (a contig, Section 2.3.5) stretching for 240 kb distal and 40 kb proximal to the probe at *D9S122* (Figure 8.8). It soon transpired that polymorphic markers obtained from the most distal clones in this walk were those (mentioned above) which showed the highest levels of linkage disequilibrium with *CF*.

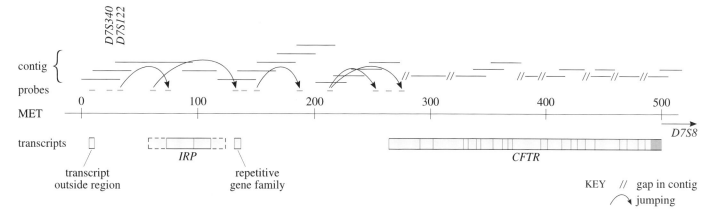

Figure 8.8 A map of the walk to the cystic fibrosis gene. The walk was initiated bidirectionally. It progressed for 230 kb before the first exon of what ultimately proved to be the gene responsible for cystic fibrosis was encountered. The cDNAs isolated with this exonic probe were used to extend the genomic walk, downstream exons providing new points of entry into genomic libraries. (N.B. the dotted lines on *IRP* indicate that its boundary is not known.)

8.2.5 Identifying genes in the interval

As the walk progressed those executing it examined the genomic clones for evidence that they contained genes.

▷ What evidence might they have looked for?

▶ Sequence conservation, CpG islands and computer analyses.

A gene is not able to accumulate mutations freely because most will be deleterious and will be removed by natural selection. Hence gene sequences are conserved. This means that probes made from the genes of one species will usually be able to form complementary base paired structures (hybridize to) genes from another species. Admittedly, the pairing will not be perfect but it will be good enough to be detectable on a Southern blot. So one approach is to probe a Southern blot containing digests of DNA from many species (a zoo blot, Section 2.6.1) with the DNA fragment to be tested hoping to see signals in more than just the human DNA.

CpG islands are readily identified by their high content of sites for rare cutting restriction enzymes and are often present at the 5′ ends of genes. There is one in *IRP* and another at the 5′ end of the *CFTR* (see below) gene.

Computer programs have been developed to look for open reading frames (Section 2.6.2) in the correct context of signal sequences (such as intron–exon junction sequences).

Four potentially transcribed sequences were found by screening zoo blots. The first was proximal to *D7S340* and was thereby excluded as a candidate, the second was *IRP*, the gene described above and already excluded, the third belonged to a transcribed repetitive DNA family (thousands of members of which existed elsewhere in the genome), and the fourth sequence, despite hybridizing strongly with bovine DNA, at first did not seem to be transcribed at all. Finally, after determined efforts, a single 920 bp cDNA clone was isolated (using this sequence as a probe) from a library made from cultured sweat gland epithelium. This cDNA, which hybridized well to a 6.5 kb mRNA on northern blots, was used to isolate further, more extensive cDNAs (despite many technical problems such as the extreme reluctance of *E. coli* to grow when containing this gene's cDNA).

The cDNAs were used to isolate corresponding genomic clones which extended the genomic walk (Figure 8.6). The gene was discovered to be approximately 230 kb in length with 27 exons. The intron–exon boundaries were discovered by comparison of the cDNA sequence (exons only) with the sequences of corresponding genomic clones (exons and introns) and the 5′ and 3′ ends were determined by direct sequencing from the mRNA. The complete sequence was 6 129 bases long with an open reading frame capable of encoding a polypeptide of 1 480 amino acids. The gene was expressed in appropriate tissues (especially in pancreas and in epithelium cultured from nasal polyps) and not in inappropriate ones such as brain tissue.

8.2.6 Identifying mutations in genes isolated from patients

Two new cDNAs were isolated from a cDNA library constructed from sweat gland epithelium cultured from a CF patient. Together they spanned the entire mRNA. Sequencing the clones revealed a 3 bp deletion removing amino acid 508 (a phenylalanine) from the polypeptide. This mutation, named ΔF508, was, it was discovered after more extensive analysis, carried by 68% of mutant chromosomes in Western Europe.

▷ What control would you use to show that the ΔF508 mutation was not a normal variant of a gene mapping close to *CF* ?

▶ It was not present in any chromosomes *known* to carry a non-mutant *CF* gene. In other words, it was only present on chromosomes known to carry a mutant *CF* gene. Because heterozygous carriers are so frequent in the population, it was not possible to use randomly chosen non-affected individuals for this important control. In fact, the chromosomes had to be proved not to carry CF by being isolated from known, non-affected heterozygotes (who could be diagnosed as so being by having had an affected child). By sequencing the gene isolated from both the normal and the affected chromosomes it was shown that that the ΔF508 mutation was indeed specific to those chromosomes with the *CF* gene.

Since then, more than 200 different mutations have been found in the gene (see below). The gene was named *CFTR*, *cystic fibrosis transmembrane conductance regulator*, in the light of its probable structure and role.

8.2.7 Deducing the nature of the CFTR gene product from the cDNA sequence

Given the DNA sequence and knowledge of the genetic code it was possible to infer the amino acid sequence of the protein produced by *CFTR*. The protein was predicted to have an internally duplicated structure. Each half comprises six hydrophobic regions, each long enough to span a cell membrane, and a domain which resembles that found in other proteins in which it is known to bind ATP. It looks similar to a well known protein called P glycoprotein which is notorious for hindering chemotherapy by transporting drugs out of cells. It differs from it by having an additional domain which contains many charged amino acids and has many sites for phosphorylation by the cAMP-dependent protein kinase. Structure alone is not sufficient to prove any protein's role. However, the repeated folding at just the scale needed to cross a cell membrane suggested that it could form a transmembrane channel. The presence of the nucleotide binding domains suggested either that a nucleotide was involved in regulating opening or closing of the channel or that a nucleotide triphosphate was used to provide a source of energy to transport something across the membrane. Thus the molecule probably functioned either as an ATP-regulated chloride channel or as an ATP-dependent transporter of chloride ions (Figure 8.9).

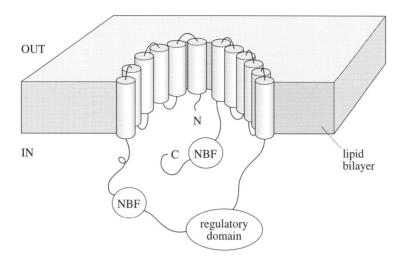

Figure 8.9 Schematic drawing showing the predicted structure of the CFTR protein (opened out). One half of the protein, represented by six membrane-spanning hydrophobic regions, links to a nucleotide binding fold (NBF), another NBF and a further group of six transmembrane domains. The common mutation, ΔF508, occurs in the first NBF. N = N terminal, C = C terminal.

Eventually, it became possible to construct a full length cDNA clone which could be propagated in *E. coli*.

▷ Why was this difficult?

▶ The cDNA was lethal to its host (see above) when expressed in *E. coli*, even if it was placed in a cloning vector that did not have a prokaryotic promoter and was not intended for expression of the gene therefore giving only very low levels of gene transcription.

▷ What method could be used to overcome this problem?

▶ A cloning vector was chosen which was maintained at only one copy per bacterial cell instead of the more usual high copy number (50–100 copies per cell). This reduced the amount of expression proportionately.

When the gene was introduced into cells in which it would not normally be expressed [such as mouse fibroblasts or HeLa (an immortal cell line derived many years ago from the breast cancer of a patient, **He**len **La**ne, who is long dead, killed by the cancer; the line is used as a standard reference cell line in laboratories all over the world)], a new cAMP-regulated chloride channel appeared. These experiments, as well as proving the nature of the CFTR protein, have opened up the possibility of **gene therapy** by proving that it is possible to obtain active CFTR protein in cells treated with the gene.

8.3   Population studies

More than 30 000 mutant chromosomes from families with *CF* have now been examined in various parts of the world and more than 230 distinct *CF* mutations described. Two-thirds of the mutant chromosomes carry the ΔF508 mutation although its frequency varies considerably in different populations. It is much more common in Britain and Scandinavia than it is in Turkey and Israel. What is more, its association with a specific haplotype is stronger in north-western Europe than in the Middle East. Perhaps ΔF508 arose in the Middle East and arrived later in the Northwest by migration. Having been present in north-western Europe for a shorter time the association with a specific haplotype is still strong compared with the somewhat weaker association in the Middle East where the original haplotype has had longer to be broken up by recombination.

Almost all *CFTR* mutations are either point mutations or very small deletions (like ΔF508). Large deletions spanning one or more exons are unusual. Why should the ΔF508 mutation have become so common in north-west Europe? Possibly heterozygotes were once at an advantage by being more resistant to the dehydration effects of infection by enterotoxic forms of *E. coli,* which cause diarrhoea and vomiting. However, direct evidence in support of this hypothesis is lacking.

8.3.1 *Carrier detection and prenatal screening*

▷ How can mutant alleles be detected in the kind of simple, quick, robust assay which is suitable for hospital laboratory use?

▶ One method would be by designing a polymerase chain reaction which will only amplify from a mutant allele. This can be accomplished, for example, by making the 3' nucleotide of one of the oligonucleotide primers correspond to a nucleotide that differs from that in the corresponding position in the normal allele. Each mutant allele would require its own specific PCR assay.

The wide variety of *CFTR* alleles makes testing for carrier status an uncertain operation. No more than about 85% of carriers in any population can be readily identified because most mutations are extremely uncommon in the population and it is economically impractical to test healthy individuals for every known *CFTR* mutation. If all carriers could be detected, partnerships that were at risk of producing affected offspring could receive counselling and prenatal diagnosis. This itself, given that one person in 22 is a carrier, raises practical and ethical problems. However for *CF* only two-thirds of carriers are easily identifiable by testing for the common allele.

▷ What proportion of at-risk couples can readily be detected?

▶ Less than half (2/3 x 2/3 = 4/9) of at-risk partnerships are detected.

This raises problems. Without careful counselling some couples may believe their children not to be at risk when the opposite is true.

Suppose that your brother or sister has cystic fibrosis and that you are not affected yourself. Straightforward Mendelian logic tells you that you have a 2/3 chance of being a carrier. You might think it wise to be tested for your status and to ask your partner to be tested as well. By studying the segregation of linked markers in your immediate family it could be deduced whether you carry the mutant gene. But what about your partner who has no previous family history of the disease? In his or her case it is now possible to give about an 85% probability of reassurance; not certainty, but at least an improvement on complete ignorance.

Perhaps you both turn out to be carriers. In this case, since your disease alleles have now been identified, it is possible to offer prenatal diagnosis of the disease status of an embryo. However, the only 'treatment' currently available is termination of pregnancy. Since the life expectancy of a cystic fibrosis sufferer has been improved by conventional medicine and physiotherapy by up to 30 to 40 years, whether or not to ask for the test and accept the termination is not an easy one to make.

8.4 Towards a cure

The gut and pancreas problems of cystic fibrosis patients can be treated with success by adding digestive enzymes to the diet. However, most patients die in their twenties and thirties because of lung damage and infection. Lung transplants provide one solution to this problem but at enormous cost and a certain risk. What might the new insight into the CFTR protein's structure and its mutations say about prospects for drug therapy?

Now that CFTR is known to be a chloride ion channel there have been many electrophysiological experiments on epithelial cells. They have shown that there is a second, previously unknown, chloride ion channel in some cells (including those of the lung). This alternative channel can be activated by raising intracellular calcium levels. Hence one possible treatment for cystic fibrosis might be to activate this channel by using a drug that lets calcium pass through cell membranes. Before testing this idea on patients, it must first be tried out by administering the drug to cells in culture and secondly by treating animals suffering from the cystic fibrosis disease. Such an untargeted attack on the problem might have severe side effects. Activation of the calcium-responsive chloride channel might not be a good thing in all tissues. An alternative possibility is gene therapy, replacement of the defective gene with a working copy.

8.4.1 Gene therapy

The words "gene therapy" have taken on a nightmarish, *Brave New World* association. Yet somatic gene therapy for inherited disorders raises no new ethical issues. If it were possible (which it is not) to treat lung epithelium with a new drug that would make defective CFTR protein work, there would be no question about ethics. As the drug does not exist, it might be possible to purify normal CFTR protein from human cells in culture and introduce it into lung cells. Still there would be no major ethical problem as this is equivalent, for example, to treating haemophiliacs with Factor VIII protein (Chapter 4). However, for cystic fibrosis the protein is unlikely to function after transfer. An attempt might be made to purify normal CFTR mRNA and add it to lung cells. There it might be translated, thus providing a working protein. Few people would worry about the ethics of this process. Unfortunately, as RNA is easily degraded it would not survive the journey into the cells. Why not, then, produce it within the cells by transcribing an introduced gene – in other words, by gene therapy? Argued in this way, there seem to be few ethical problems in introducing genes into somatic cells. However, it is important to acknowledge the caution felt by the public in interfering with what seems a fundamental attribute of life. It is also the case that, now and for the foreseeable future, neither lawyers nor geneticists are willing to accept gene therapy in human germinal, as opposed to somatic, cells because of the obvious consequences in terms of altering inheritance. In any case, before gene therapy becomes a standard treatment there remain a number of scientific obstacles to be overcome.

8.4.2 Constructing a gene for therapy

First, a working gene is needed. It would be pointless to attempt gene therapy with a replacement gene which was itself defective. However, the full-length *CFTR* gene is more than 230 kb long; far too long to be routinely cloned in bacteria using today's techniques and difficult to purify in large amounts. In addition, it would be introduced into recipient human cells only very inefficiently and with a high probability of undergoing rearrangement. Instead, a small intronless gene has been synthesized. This comprises the coding region of the cDNA of *CFTR* with added promoter and poly A addition signals (Chapter 3) taken from another widely expressed and well-studied gene.

8.4.3 Introducing the construct into cells

This gene must be inserted into cells efficiently enough so that it is expressed at a useful level in an adequate proportion of them. Luckily, epithelial cells are joined by gap junctions (Book 3, Chapter 4). If in just 10% of the cells the *CFTR* gene is expressed, then mRNA will be transported throughout the tissue via the junctions and the CFTR protein will be made in all of them. There are two main ways of getting a gene into living cells. One is to insert it into the genome of a virus able to infect the cells. If the virus has had any harmful genes removed, this should cause no problems and might be the most efficient method of gene transfer. However, there exist within our genomes the remains of many viruses which have entered in the past and have been inactivated by mutation. Perhaps introducing a living, albeit defective, virus into the cell may, by complementing the genetic defect of an endogenous virus, reactivate it. Functioning viruses could then be produced by recombination between the two strains and an infection triggered. Many geneticists feel that on the whole it is probably safer to take an alternative (but less efficient) route: to introduce the working gene wrapped up in an artificial lipid coat, a *liposome*. Liposomes containing DNA are made by vigorous mixing of lipid and DNA solution. An emulsion of tiny lipid bilayer spheres containing the DNA solution is formed. When mixed with cells the spheres merge with the cell membranes releasing their contents into the cell interiors.

8.4.4 Assessing the effectiveness of treatment

Gene therapy for cystic fibrosis, and indeed for any other condition, involved a number of trials. First the replacement gene was tested on cultures of epithelial cells derived from CF patients. Could it restore the missing ion channel? It could. Next the construct and its liposome delivery system were tested on normal animals. Was it toxic? It was not. Then the gene was used to create transgenic mice, animals in which the human *CFTR* gene had been inserted into the germ-line. Because the gene had inserted at random into the transgenic animal's genome it was not under normal control and consequently these animals expressed the CFTR protein in many unusual sites in their bodies. This might also happen if human patients were given the gene. Was this over-expression harmful? It did not seem to be. The transgenic mice expressing the human *CFTR* gene at a high level suffer no apparent ill effects. There remained, though, the central question. Could it cure the disease? In order to answer this, further research required a mutant mouse with cystic fibrosis.

8.4.5 A mouse model of the disease

The cystic fibrosis mutation has never been found in mice. An 'artificial mouse' had to be made and several laboratories succeeded in *knocking out*, that is, deleting, the mouse *cftr* gene. Although the methods differed in detail, in general the approach was the same (Figure 8.10). In each case a mutant gene was constructed *in vitro* and then introduced into cultured **embryonic stem cells** (ES cells), originally derived from a mouse embryo. Such cells can be reintro- duced into an early embryo where they may contribute to the resulting fetus. After DNA enters an ES cell it may integrate into the genome. Rarely (but often enough to make the experiment feasible) the introduced mutant *cftr* gene

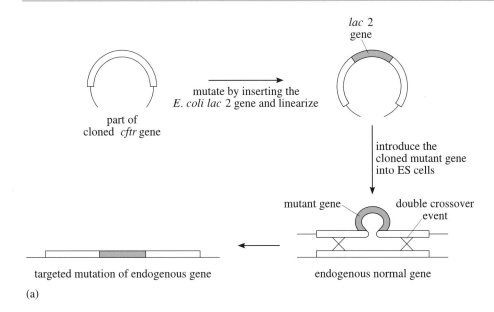

(a)

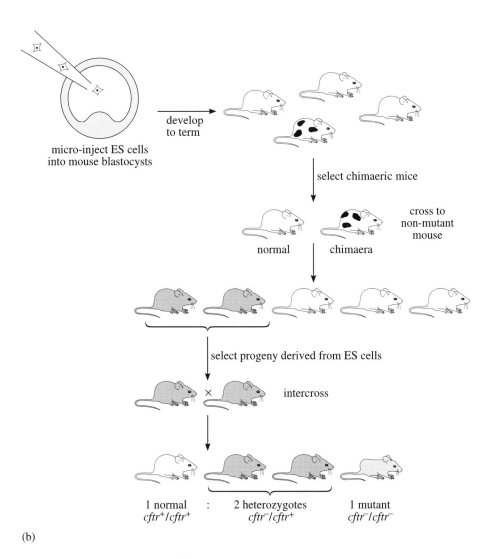

(b)

Figure 8.10 The experimental stages involved in the development of a mouse with a mutant *cftr* gene. (a) *In vitro* construction of the mutant gene, insertion into ES cells and integration into the endogenous normal *cftr* gene. (b) Creation of chimaeras and production of homozygous mutant offspring by selective breeding.

integrated by recombining into the endogenous normal *cftr* gene, inactivating it. Cells descended from such an event were inserted into early mouse embryos creating **chimaeras** (individuals which are a mixture of two distinct cell lineages) and these were re-implanted into female mice to continue development. Sometimes the ES cells contributed to the cells that were destined to form sperm or egg cells in the chimaeras. These mice could be mated to normal mice to produce progeny heterozygous for the introduced *cftr* mutation. Simple crossing of the heterozygotes was then carried out to produce homozygous mutant offspring that should potentially be suffering from a novel condition in mice, cystic fibrosis.

Different lines of 'transgenic knockout' mice suffer from cystic fibrosis with varying degrees of severity. Some die shortly after birth with severe intestinal blockage resembling the blocked bowel found in some human new-borns with cystic fibrosis. In other strains the pancreatic duct is blocked in older mice but not in young. No severe lung problems were seen in any of the knockout mice but perhaps this is because the mice were raised in a pathogen-free environment and lung infection is required before any pathological changes are observed.

Crucially, the chloride channels in the knockout mice were not functioning properly. This damage did, it transpired, respond to gene therapy. A nebulizer (which produces a fine mist of the liposome emulsion) was used to introduce a copy of the working gene in its liposome coat into the lungs. After treatment functioning chloride channels were found. As many as 70% of the animals responded to therapy albeit with a wide range of levels of expression of the chloride channel.

8.4.6   A clinical trial

The results with mice were encouraging enough for a clinical trial to have begun with cystic fibrosis patients. It was initially thought to be too difficult to introduce the gene into lung epithelium. Instead the epithelial cells lining the nose were studied. The results are not yet published at the time of writing.

In spite of the potential of gene therapy, it is unlikely to effect a permanent cure. The cells lining the lung (which receive the *CFTR* gene) do not divide and have a lifetime of about 60 days. The treatment will probably have to be repeated at about this interval. Ultimately a gene therapy may emerge that places the working gene into the stem cells which replenish the epithelial population thus giving an effective cure. There is, though, the important problem that it is difficult to target these cells directly and they exist in relatively small numbers.

Even if gene therapy is able to cut the need for painful physiotherapy by only 50%, it might be of benefit. Whether it is commercially viable, however, is less certain. Perhaps one in 2 500 of the population is not a sufficient market, although each patient would require treatment for their lifetime.

Summary of Chapter 8

Human genetics has gone through two distinct stages since the molecular revolution of the 1970s and 1980s and is now again approaching another transition. In the first stage, the new cloning technology was applied to biological and biochemical information discovered in the 1960s. In this way the genes coding for many well known proteins, such as the α- and β-globins or the various blood clotting factors, were identified. The second stage, exemplified by the story told here of the cloning of the cystic fibrosis gene, was of cloning well-known disease genes of unknown biochemistry or function. The stage which is just beginning is of the investigation of the thousands of genes of unknown function which are being revealed by their proximity in the genome to genes such as *CFTR* and by the Human Genome Project (Video sequence 4).

Cloning cystic fibrosis involved scientists from many laboratories in a bitterly competitive race. Ultimately it was successful as far as one laboratory was concerned. One lesson learnt from this has been the benefit of cooperation rather than competition and it is noteworthy that a number of reports of gene cloning have recently been authored by consortia containing representatives of many laboratories with no one laboratory given prominence. The cystic fibrosis story, as well as being a paradigm of a positional cloning success story, has also moved on to what was, not long ago, science fiction, that is, the creation of designer mutations in mice and of gene therapy in both humans and mouse. In this it is again in the forefront of a fast moving field and it will be fascinating to observe what addenda are added to this chapter over the next five years.

Objectives for Chapter 8

8.1   Define and use, or recognize definitions and applications of, each of the terms printed in **bold** in the text.

8.2   By use of the *CF* gene as a paradigm of positional cloning, demonstrate how evidence from a variety of experiments ranging from DNA sequencing to the study of population genetics contributed to identifying the *CFTR* gene.

8.3   Using cystic fibrosis as a model, discuss the concept and problems associated with screening a population for a genetic disease.

8.4   Using cystic fibrosis as a model, discuss the steps required to validate a potential gene therapy.

Questions for Chapter 8

Question 8.1   (Objective 8.2)

Explain why the term 'reverse genetics' at first seemed an appropriate description of the process now known as positional cloning.

Question 8.2   (Objective 8.2)

Explain how chromosome jumping overcomes the problems associated with chromosome walking.

Question 8.3   (Objective 8.3)

If there existed a test which could identify more than 99% of carriers of mutant *CFTR* genes, as is the case for some other genetic diseases such as fragile-X syndrome or Huntington's disease, what would be the arguments for and against screening the entire population? How is this affected by our inability to detect (in a routine assay) more than about 85% of carrier individuals?

Question 8.4   (Objective 8.4)

What benefits come from making mouse models of human genetic diseases?

Answers to questions

Question 1.1

The green plants are all homozygous for the green pea colour allele ($y\,y$); but the yellows are of two different genotypes. Half the total progeny (that is, two-thirds of the yellows) are heterozygotes ($Y\,y$), producing both greens and yellows when selfed, and a quarter of the total progeny (one third of the yellows) are homozygotes ($Y\,Y$).

Question 1.2

(a) All the daughters of an affected male are affected (as they all receive his X chromosome with its single mutant allele). Half the sons and half the daughters of an affected female are themselves affected.

(b) II-2 is a heterozygous female ($X^{Pd}X$), II-4 is an unaffected male hemizygous for the normal allele (XY) and III-5 is a male hemizygous for the phosphate diabetes allele ($X^{Pd}Y$).

Question 1.3

Every child born with the disease has inherited two copies of the damaged gene, one from each parent. If, in the land of Albany, one fertilization in 3 600 has a copy in both the sperm and the egg that take part in it, one Albanian sperm and one Albanian egg in 60 must carry the gene. This means that 59 out of 60 do *not* carry it. This, in turn, means that the vast majority of fertilizations involving the damaged skin-pigment gene involve one damaged copy and one normal copy, giving rise, as it is recessive, to perfectly normal children. There are far more copies of the damaged gene hidden in heterozygotes than there are in affected homozygotes. Preventing homozygotes from reproducing will therefore lead to only a very small decrease in the number of affected children in the next generation: the dictator does not understand Mendelism and his policy will fail.

The Hardy–Weinberg formula (Box 1.1) allows the frequencies of the various genotypes in cases such as these to be calculated exactly.

Question 2.1

Aneuploidy means that an individual has an abnormal number of chromosomes, that is, more or less than 46. Excess is usually tolerated better than deficiency, so trisomy (three copies) of chromosomes 13, 18 and 21 are viable but the only viable monosomy (one copy) is the presence of one copy of the X chromosome. Sex chromosome aneuploidies exist; for example Klinefelter's syndrome individuals (XXY) are male, usually sterile and have some degree of mental retardation. Females with up to five X chromosomes have been described but these are usually normal (apart from some mental retardation) due to the phenomenon of X inactivation (Chapter 3).

The karyotype in Figure 2.35 is that of a person with Down's syndrome (trisomy 21). These individuals have a range of phenotypic features: they have varying degrees of mental retardation, short stature and a characteristic small round head and broad hands and feet. They may be subject to heart and eye problems and have an increased risk of developing leukaemia.

Question 2.2

(a) The gene could be assigned to a chromosome using somatic cell hybrids. The enzyme activity could be tested for in cellular extracts from a panel of hybrids. Those in which the enzyme activity is detected are the ones that contain the human chromosome on which the gene resides. There are some problems with the technique: it is possible that the human gene may not be expressed in the hybrids and it is also possible that the human enzyme activity test will cross-react with similar rodent enzymes.

(b) It would be necessary to carry out linkage analysis. DNA from as many individuals as possible in the families would be extracted and tested with a series of markers from around the genome. First, the affected parent would be tested to see whether they were heterozygous at the disease locus. Selected markers would then be used to test affected and unaffected offspring to see whether the alleles of the marker co-segregate with the disease (think back to the ABO blood group markers and the nail–patella syndrome). Once a linked marker is found, it is necessary to test further markers in the region in order to pinpoint the position of the gene as much as possible. When close-flanking markers have been found, the region between them can be isolated by physical mapping techniques.

(c) Since the cloned region is deleted in all the individuals, it is likely that the disease gene is present in that region. A good way to start looking for genes would be to try and find CpG islands by digesting clones in the region with rare cutter restriction enzymes. Alternatively, short fragments from the clones could be tested for conservation or expression by hybridization to zoo blots or Northern blots respectively. If the region was small it might be possible to sequence it and then look for sequences that are characteristic of genes such as open reading frames or conserved sequence motifs.

Question 2.3

A contig is a series of continuous overlapping DNA fragments. The term was originally coined in reference to overlapping portions of nucleotide sequence but is also often used to refer to any overlapping cloned fragments. A contig is representative of a much larger region of DNA. When looking for genes or beginning to sequence a chromosomal region, the set of small clones provides a manageable starting point.

Question 2.4

It is not known exactly how many genes there are in the human genome but estimates range from 30 000 to 100 000. Although this seems a lot, the amount of DNA needed to encode the proteins is only around 5% of the genome. This is in marked contrast to the genome of phage ϕX174 where 96% of the genome is coding. Eukaryotic genes contain about ten times more DNA than they need for protein coding; this extra DNA takes the form of introns and regulatory regions. It is possible that the introns contain some information necessary for gene regulation. Some genes are present more than once – they are members of a gene family. Examples include the haemoglobin genes and those encoding tRNAs and rRNAs. These additional copies may be important when a lot of gene product is required. Some members of gene families are no longer coding – they are pseudogenes. Although mutations in extra gene copies may cause them to lose their function, it is also possible that some changes will be advantageous and will be selected for, perhaps leading to a new role.

Up to 30% of the human genome appears to be made up of repetitive sequences. Some of these are repeated many times in large tandem arrays which seem to be much reduced in size in some individuals without any apparent effect. It is very difficult, therefore, to assign a function to these regions. However, there is evidence that tandem repeats at the telomeres and centromeres are important. The telomeric repeats stop the chromosome 'fraying' at the ends and the centromeric repeats may be important in the alignment of homologous chromosomes and spindle attachment during cell division. It is possible that, with further study, the functions of other types of 'junk' DNA may be identified.

Question 2.5

(a) The aims of the Human Genome Project are to map, sequence and analyse the complete human genome; thus identifying all of the genes and working out what they all do. This will be the starting point for the treatment of many genetic conditions. For example, the gene therapy trials that aim to treat cystic fibrosis (Chapter 8) could not have been attempted without knowledge of the DNA sequence of the gene. Such knowledge is also important when screening for genetic conditions. Once the mutation within a gene that causes a disease is known, it is potentially possible to screen individuals in a population, and maybe to identify carriers. This has been done to identify carriers of the mutated allele of the Tay–Sachs gene, carriers can then avoid marrying each other. Genetic screening is also important for conditions where alteration in lifestyle early on in life could go some way towards ameliorating the disease symptoms. For example, in the UK all new-born babies are tested for phenylalanine hydroxylase activity in a small blood sample. Lack of this enzyme leads to a build up of the amino acid phenylalanine, which causes mental retardation. However, when detected early, children who are homozygous for the mutant allele and so do not have sufficient enzyme activity can be placed on low-phenylalanine diets. Using the knowledge of gene sequences obtained from the Human Genome Project it is potentially feasible to screen for a large number of genetic conditions. However, this is more difficult than it may seem, as several mutations in the *same* gene may cause disease, and so for each condition several mutations will have to be screened for (as is the case for the cystic fibrosis gene – Chapter 8).

If we learn what enzyme or hormone is deficient in a genetic condition, then it may be possible to administer that protein; for example, insulin to diabetics. By working out what the gene products are and what they do we might come across more conditions that can be treated in this way.

(b) The money invested in the Human Genome Project in the USA has come predominantly from defence funds that are no longer required. It seems likely that much medical benefit will come from the project. However, many argue that the money would be much better spent treating non-genetic disease in the developing world which affects many more people and can be easily treated by provision of clean water supplies, vaccination and basic medical care. In the developed world, much money is spent caring for people suffering from conditions which could possibly be prevented by the use of the knowledge provided by the Human Genome Project.

(c) There are objections to the Human Genome Project on ethical grounds. In fact, more than 5% of the budget has been put aside to be spent on considering these issues and producing policies so that the public can be assured that the data will be used in the best possible way. A major objection is on the grounds of privacy. If it is a theoretical possibility to know the whole of one's genetic future and this is stored in a database, then it is likely the data will be accessible by other parties. This may have an effect on personal health insurance – perhaps increased premiums for people likely to develop a genetic disease. It may also adversely affect employment prospects – people who are carriers for sickle-cell disease have already been stigmatized and discriminated against in the job market even though they are perfectly healthy.

In terms of screening, it seems that there is little point in screening for conditions where no treatment is available. This is particularly true in the case of late-onset diseases such as Huntington's disease, where a person may acquire the knowledge that they will develop a seriously debilitating condition when they reach middle age. However, some people would rather know than live under the shadow of doubt and the knowledge may help them to plan whether they have children or not as they risk passing this dominant allele to their offspring. In general, the gathering of the data doesn't pose an ethical problem (apart from the costs involved) but it is the use of this data when it becomes available for individuals that must be strictly controlled.

These topics are discussed in more detail in Video sequence 4.

Question 3.1

The messenger RNA (mRNA) is the template for translation; it carries the genetic code. The ribosomal RNA (rRNA) is a constituent of the ribosome which contains proteins and three types of RNA (28S, 18S and 5S). The ribosomal RNA (28S and 18S) is produced from clusters of gene families on the short arms of the acrocentric chromosomes (Chapter 2). (The genes for 5S rRNA are on chromosome 1.) The RNA molecules are incorporated into the ribosome when the nucleolar organizer regions of the acrocentric chromosomes come together to form the nucleoli. The role of RNA in the ribosome is not certain, but it is likely to have a catalytic role and also to be involved in the recognition and binding of the mRNA.

Transfer RNA (tRNA) molecules bring the amino acids to join the growing peptide chain. They consist of an anticodon at one end which pairs with the codons of the mRNA. This ensures that the correct amino acid is brought to the site. At their opposite end they carry the amino acid. There are more tRNA species than there are amino acids as each amino acid can be specified by more than one codon (third base degeneracy or 'wobble').

Question 3.2

Pre-mRNA contains exons and introns, as well as the 5′ leader and 3′ trailer sequences. It is processed in several ways to form mRNA which is exported from the nucleus. The introns are removed by the process of RNA splicing, a 5′ cap (7-methyl guanosine triphosphate) is added, and this is recognized by the ribosome. A polyadenosine tail is added at the 3′ end, and this probably enhances mRNA stability.

Question 3.3

The three genes involved in lactose metabolism, *lac Z, lac Y* and *lac A,* are expressed in concert as a polycistronic mRNA in response to the presence of lactose in the medium surrounding the cell. If lactose is not present, the genes are not expressed. This is due to the action of the *lac I* gene product which binds as a tetramer to the operator of the *lac* operon and prevents transcription taking place. When lactose is present, some molecules are converted to allolactose by basal levels of β-galactosidase. The allolactose binds to the repressor which changes conformation and can no longer can bind the operator. Removal of the repressor molecule allows transcription to take place. This is an example of positive regulation: the genes are not expressed until an inducer (allolactose) is present. Some operons are under negative control, the gene products are constitutively made until a repressor molecule – maybe the end product of a metabolic pathway – shuts off transcription.

Question 3.4

Prokaryotic genes do not contain introns; in addition, often only a very small proportion of the genome is non-coding, to the extent that groups of genes may be arranged sequentially and transcribed into a single polycistronic mRNA molecule. Regulatory genes may also be located nearby and act in *cis* mode (that is, genes on the same strand controlling expression of others on that strand) to control gene expression. The promoter region consists of a group of conserved DNA sequence motifs and determines where transcription is initiated. Transcription is carried out by a single RNA polymerase and proceeds until the termination sequence is reached.

Eukaryotic genes contain a much greater proportion of non-coding sequences. These include introns which are spliced out of the RNA before it is translated and also other regulatory regions. Some of these regulatory regions are part of the transcription unit; these include the 5′ leader and the 3′ trailer, which remain in the mRNA but are not translated. The distance between one eukaryotic gene and the next may be very great, and sequences that are involved in regulation of the genes may be positioned some distance away from the gene upon which they have their effect. Eukaryotic genes are transcribed by three different RNA polymerases and several proteins are involved in the recognition of the promoter and the formation of a transcription complex. Transcription is regulated by *trans* acting transcription factors, some of which are ubiquitous and others tissue-specific.

Question 3.5

There are several levels at which transcription can be regulated in eukaryotes. First, due to the multi-unit RNA polymerase transcription complex that must be built up and the activation of this complex by other factors, which in turn may become active only in the presence of a cofactor, it is apparent that the presence or absence of these factors will allow sophisticated control. Thus, genes may contain within their promoters or regulatory region, DNA sequence elements which will allow the gene to be expressed in a range of tissues under different circumstances (for example, the heat shock response elements and enhancer and silencer sequences).

Transcriptional control is also brought about by chromatin structure. Genes that are capable of being transcribed in any one tissue have an open chromatin structure in that tissue, whereas those that are never expressed there will be packaged as condensed chromatin. Regions that don't seem to be expressed at all appear to be packed as heterochromatin. The open chromatin structure may allow access to the DNA by transcription factors and other regulatory molecules such as steroid hormone receptor complexes.

DNA methylation is also involved in determining which genes are expressed in different tissues, at least in mammals. Certain proteins which bind to highly methylated promoter sequences may block access by transcription factors.

Question 3.6

Transcription factors are regulatory proteins which bind specifically to DNA sequence motifs within promoters or enhancers and which have an effect on gene expression. Since different transcription factors are present in different tissues this allows different genes to be expressed and differentiation to occur.

Since transcription factors must recognize particular elements within the DNA, they must contain a DNA-binding region. In addition, they possess a region which interacts with the RNA transcription complex and promotes gene expression. Several different classes of transcription factors have been described on the basis of their protein structure. DNA binding domains include the helix-turn-helix, the zinc finger and the leucine zipper. Each of these conformations allows specific sequences within the DNA to be recognized and bound.

Question 3.7

Protein	transferrin receptor (TfR)	ferritin
Function	uptake of iron into cells	iron storage
Mechanism of control	3′ untranslated region forms five stem-loop structures which bind iron responsive element binding protein (IRE-BP) when iron is scarce	5′ untranslated region forms stem-loop structure, also binds IRE-BP when iron is scarce
Iron abundant	*reduced* TfR, IRE-BP not bound, mRNA degraded	*increased* ferritin, IRE-BP not bound, ferritin mRNA translated
Iron scarce	*increased* TfR, IRE-BP binds to stem loops, stops mRNA degradation	*decreased* ferritin, IRE-BP bound to 5′ end of mRNA, and stops translation

Question 3.8

As a means of dosage compensation in the gene expression between males and females, one X chromosome is inactivated (switched off) in females during early embryonic development. This is a random and permanent process through mitotic cell division. Thus in different cells the female may have either her paternal or her maternal X chromosome inactivated. This means that females are effectively mosaics, as they will inevitably inherit different alleles at loci on the maternal and paternal X chromosomes. The inactive X appears as very dense heterochromatin and very few genes are expressed from it. In the case of the tortoise-shell cat, an X-linked allele for orange hair (*O*) is dominant to the allele (*o*) for black hair. If X inactivation did not occur, a heterozygous cat (*O o*) would have orange fur. However, where clones of cells arise that contain an inactivated *O* allele, the hairs will be black. Where the *o* allele is inactivated the hairs will be orange. Males would have fur that was either black or orange all over, depending which allele they carried at this locus on their single X chromosome (also the pattern and colour varies due to alleles at other loci).

Question 3.9

The hormone insulin regulates the uptake of glucose into cells and its conversion to glycogen; it also stimulates protein and lipid synthesis.

One cause of diabetes is overeating. When there is constant glucose in the bloodstream the cells cannot produce enough insulin receptors, and so the cells become insensitive to the hormone; this usually results in obesity. However, in some individuals, who for some reason cannot keep producing enough insulin to direct the cells to take up all the glucose, the glucose remains in the bloodstream and leads to the symptoms of diabetes.

Some individuals suffer from insulin-dependent diabetes (IDDM). In this auto-immune condition the β cells of the pancreas, which normally produce insulin, are attacked. There is thus a lack of insulin, which leads to diabetes. The condition can be treated by daily administration of insulin.

Non-insulin-dependent diabetes (NIDDM) may also be linked to obesity. In this case, although insulin is present, there seems to be an insensitivity to insulin in the peripheral tissues. This can be due to several factors:

o mutations that affect the processing of the insulin precursor protein;

o mutations that alter particular amino acids and so affect the function of the insulin protein;

o lack or depletion of insulin receptors;

o other defects in the insulin signalling pathway.

The symptoms of NIDDM can be treated by an appropriate diet or agents that increase the transcription of genes encoding glucose transporter proteins – thereby bypassing the insulin signalling pathway.

Question 4.1

In the early days of genetic study, only dominant mutations with large effects were detected – think of Darwin and his 'sports'. They were thought to be rare and even that they might lead to the origin of new species. Methods for uncovering recessives made it possible to look at the effects of mutagens. As most mutations were then still seen as harmful there was concern that a slight increase in, for example, radiation would cause permanent damage. The advent of protein electrophoresis was followed by the development of restriction enzyme techniques, research tools that began to reveal how populations are full of diversity (all of which must have arisen through mutation). Most such mutations seemed to have no effect and so these days, most are regarded as harmless. The new methods of DNA sequencing showed that what might seem like a simple mutation with straightforward effects might, like the haemophilia condition, actually represent hundreds of different mutational events. It suggested that mutation has a dynamism of its own, and that segments of the DNA may undergo genetic change in their own interests rather than in those of their carriers.

Question 4.2

Each chromosome contains thousands of genes. Any change seen down the microscope has, inevitably, effects more major than those influencing a few DNA sequences. There are, nevertheless, parallels between the two. *Aneuploidy* – the gain or loss of a chromosome – is a sometimes viable, but usually lethal condition. It has no common equivalent at the DNA level. Chromosomal *inversions* – reordering of the chromosomal material – can be seen on a much smaller scale in DNA, as can *duplications*. DNA can duplicate itself not just once, but many times, leading to repeated sequences. DNA sequences can, it seems, move between chromosomes, analogous to the *translocations* seen down the microscope.

Deletions in chromosome structure give rise to visible mutations (such as the *notch* mutation in *Drosophila*); and many inborn genetic diseases are due to the deletion of short sequences of DNA (think, for example, of the thalassaemias).

Question 4.3

Certainly, humans are exposed to more radiation now than they once were; this will increase the rate of mutation. The effects of chemicals are harder to assess. Some new chemicals cause mutation, but so do hundreds of others that have been part of the environment since humans evolved. Plenty of plants (particularly those with a bitter taste that might, perhaps, be eaten only when food is in short supply) are full of mutagens and badly stored grains can be covered with fungi that produce potent and damaging chemicals. The most powerful mutagen of all is old age. Once, people reproduced until they were too old to do so; many children were born to elderly parents. This led, no doubt, to many new mutations. Now, there are fewer very old parents and fewer very young parents. On balance (and nobody really knows), the mutation rate has probably gone down.

Question 5.1

The cousins share two grandparents in common. If one of these grandparents carries a recessive deleterious allele (he or she has the genotype *Aa*) and the other does not, each of their children has a one in two chance of being a carrier. Each of those children has, if they did indeed inherit the damaged gene, a one in two chance of passing it to the next generation. Each grandchild, then, has a one in four chance of inheriting a single copy. The chance of both of them inheriting is $1/4 \times 1/4$, which is one in sixteen. If they – the cousins – marry, the chance of any child receiving two copies (genotype *aa*) and suffering genetic damage as a result is that probability multiplied by the standard Mendelian chance of a recessive homozygote emerging from a mating between two heterozygotes; that is, $1/16$ multiplied by $1/4$ or one in sixty four. This chance is the same for any locus segregating for recessive alleles reducing fitness.

If one of the polymorphisms at the DNA level had some effect on fitness – that is, was open to the the action of natural selection – then inbreeding would have a large effect on the health of the children. Its rather modest influence suggests that in fact the vast majority of such polymorphisms have no effect on the ability of their carriers to survive and reproduce – that they are neutral and are ignored by natural selection.

Question 5.2

Many things could explain the remarkably slight genetic differences between ourselves and our closest relatives. Perhaps the genes responsible for morphological evolution in primates are few in number. If, for example, one gene mutation was enough to slow down the growth rate of humans so that they become sexually mature while retaining many juvenile features, then a genetic change so small as to be virtually undetectable would have major effects on shape, size and perhaps, behaviour. The genetic changes in the snails might be easier to detect, but have no such fundamental effect. Possibly, the snails have been separated in evolutionary time for longer than have chimps and humans and have lived in such an unchanging environment that natural selection has not produced morphological change. Possibly – and there is evidence that this is true

– snail DNA is intrinsically more dynamic and unstable than is that of primates, and sequences are sweeping through it under their own volition with no effect on the animal's appearance.

Perhaps there is in fact almost no morphological difference between chimps and humans – we only think so because we are biased observers – and what separates them is cultural evolution, most notably the development of language, which leaves no traces in the DNA.

Question 6.1

Sex-linked characters are, in most cases, transmitted by mothers but show their effects more frequently in sons. Affected sons may produce daughters who can themselves transmit the character to a son. Mitochondrial mutations, though, are transmitted by mothers to both daughters and sons; but only daughters pass the mutation on to the next generation.

Question 6.2

There might be somatic variation; in other words, different tissues show the phenotype to a different extent because of the segregation of mitochondria during mitosis or the local accumulation of high frequencies of a mitochondrial mutation. The symptoms could become worse with age, as mitochondrial mutations accumulate. Any syndrome which is associated with tissues showing high levels of oxidative phosphorylation and cell respiration (such as muscle) may involve a mitochondrial change as this is where much of the cell's energy metabolism takes place.

Question 7.1

A mosaic patch may develop as a result of somatic mutation. The later in development the mutation occurs, the smaller the patch of tissue derived from that mutant cell.

Question 7.2

An immunoglobulin gene spans many millions of base pairs and is divided into *V*, *J* and *C* segments (and *D* region in the case of the heavy chain gene). The *V* segment consists of at least 50 genes. The *J* and *D* regions also contain multiple genes. During maturation, the B cell gene cluster undergoes somatic rearrangement so that a *V*, a *J* and a *C* gene are brought adjacent to each other to give a different combination from that in other developing B cells. This combination of genes is then transcribed; intervening genes and non-coding regions are spliced out.

Somatic mutation within the *V* region genes increases diversity, and yet more differences arise from the assembly of two heavy and two light chains to produce the unique antibody.

Question 7.3

In the case of CML, the Philadelphia chromosome arises as a reciprocal translocation between chromosomes 9 and 22. This results in the production of a fusion gene between *abl* and *bcr*. This fusion gene is transcribed to produce a new gene product. An alternative example is Burkitt's lymphoma. Here a reciprocal translocation between chromosomes 8 and 14 leads to enhanced expression of the *myc* oncogene.

Question 7.4

A virus carrying an oncogene introduces it into a new cell on infection. Viral oncogenes are altered versions of proto-oncogenes, differing from them in that they are often truncated and have accumulated point mutations. Oncogenes, with their changed structure and function, are dominant to normal cellular proto-oncogenes from which they are derived.

Question 7.5

Heterozygotes for *Rb* arise in two ways: individuals with the familial form inherit a mutant allele from one parent, those with the sporadic form acquire a mutant allele during development, early enough for a number of tissues to be heterozygous. Tumour-suppressor genes are recessive; a second hit is required for tumour growth. Tumour DNA from heterozygotes can reveal this. It shows that these cells are no longer heterozygous (loss of heterozygosity), either because the normal allele mutates (due to a point mutation), is lost (due to either a deletion, or loss of a whole chromosome), or is homozygous for the mutant allele (mitotic recombination, or chromosome loss followed by duplication of the surviving chromosome).

Question 7.6

You may have given any two of the following (or other examples in the text):

1   The non-linear increase in cancer with age suggests that the development of cancer involves a series of changes.

2   The presence of the Philadelphia chromosome on its own is not sufficient for CML to develop.

3   The presence of an 8 to 14 translocation is not sufficient on its own for Burkitt's lymphoma to develop.

4   The time-lag between a victim's exposure to radiation from atomic bombs and a cancer developing suggests that additional mutations accumulate.

Question 7.7

There are a number of factors, but they can be divided simply into:

1   Genetic factors [for example, fast versus slow acetylators, and polymorphism of the enzyme AHH (aryl hydrocarbon hydroxylase)].

2   Environmental factors [for example, diet (fruit and vegetable intake), *number* of cigarettes smoked per day and other environmental factors such as exposure to asbestos].

3   An interplay between 1 and 2 (for example, fast acetylators who smoke convert harmful aromatic amines into harmless products quicker than slow acetylators, so are at a decreased risk).

Question 8.1

The science of genetics has a history. The way that experiments are designed is to a large extent dependent on that history. When the techniques of gene cloning first became available it was logical to apply them to genes where the protein products, such as the α or β globins or enzymes such as α_1-antitrypsin, were already well characterized. Something about the protein, its primary structure perhaps, its tissue specific pattern of expression, or an antibody raised against it, could be used either to provide the information needed to design the cloning experiments or, in the case of an antibody, actually to select the correct clone from an expression library. The logic of the experiments went from knowledge of the protein product to cloning the gene and eventually to mapping its location in the genome. When all the easy targets had been exhausted, the positional cloning method was invented which allowed genes to be cloned from no more original knowledge than genetic mapping studies in affected families. The logic went from map position to cloning candidate genes to proof of identity by discovery of mutations in affected individuals. From the DNA sequence the primary structure of the protein product could be deduced and from that it was sometimes possible to infer the role of the protein. This logical pathway from map position to knowledge of the protein product is the precise reverse of the 'conventional' pathway.

Question 8.2

Walking is laborious and cannot easily pass 'unclonable' regions or regions containing a high concentration of repeated sequence DNA. Jumping has less problems with either of these obstacles. Unclonable regions are usually very short, they will not be present in the library but this does not matter because they will be jumped over, unlike walking where all regions need to be present in the clones of the walk. Repeated sequences can be more troublesome but if enough jumps are taken it is usually possible for at least one jump to alight beyond or between them. Walks can be initiated from all the jumping points; this greatly speeds up the walking process by giving many more ends to extend.

Question 8.3

At 99%:

In favour:

(a) The ability to identify all partnerships at risk of producing an affected child.

(b) A consequent reduction in the number of cystic fibrosis sufferers with benefits both in the net reduction of suffering (to both affected children and to their families) and in cost to the Health Service.

Against:

(a) Unnecessary alarm caused by false positives (probably not a big problem because the tests can be repeated).

(b) Mistaken sense of security caused by false negatives (hopefully there will not be too many of these).

(c) Discrimination by employers and health insurers against identified carriers (a good programme of public education is required to avoid this).

At 85%:

In favour:

(a) Ability to identify 72% of partnerships at risk of producing an affected child.

Against:

(a) A high proportion (28%) of partnerships which are at risk will not be identified. This type of probabilistic information is very difficult to explain to a couple and not very helpful. 'You have not been given clean bills of health but you are not definitely carriers. Your personal risks of being carriers have shrunk from the population mean (~1/22) to 15% of that.' etc.

Question 8.4

(a) The mutant animals provide further information about possible biochemical and physiological effects of the disease.

(b) The mutant animals provide a system on which to try the effects of new treatments.

Further reading

Chapter 1

Griffiths, A. J. G., Miller, J. H., Suzuki, D. T., Lewontin, R. C. and Gelbart, W. M. (1993) *An Introduction to Genetic Analysis*, 5th edn, W. H. Freeman and Company.

> There are many excellent textbooks of genetics and this is the most comprehensive.

Jones, S. (1994) *The Language of the Genes*, Harper Collins.

> An introduction to human genetics for the general reader.

Maxson, L. R. and Daugherty, C. (1992) *Genetics, a Human Perspective*, W. C. Brown.

> A good introduction to specifically human aspects of genetics.

Chapter 2

McConkey, E. H. (1993) *Human Genetics: The Molecular Revolution*, Jones and Bartlett.

Mange, A. P. and Mange, E. J. (1980) *Genetics: Human Aspects*, 3rd edn, Sinauer Press.

Wicking, C. and Williamson, B. (1991) From linked marker to gene, *Trends in Genetics*, **7**, No. 9, 288–292 (review article).

Chapter 3

Latchman, D. (1990) *Gene Regulation: A Eukaryotic Perspective*, Unwin Hyman.

MacLean, N. (1989) *Genes and Gene Regulation*, Edward Arnold Series – new studies in biology.

Lewin, B. (1994) *Genes V*, selected chapters: 7, 8, 9, 10, 14, 15, 28, 29, 30, 31 and 32, Oxford University Press.

Current Opinion in Genetics and Development: (1994) Vol. 4, No. 2.

> Contains a series of review articles, all of which are relevant to the topics discussed in this chapter.

Chapter 4

Collins and Weissman (1985) The molecular genetics of human haemoglobins, *Prog. Nucleic Acid Res. and Mol. Biol.*, **31**, 315–462.

Friedberg, E., Walker, G. and Siede, W. (1995) *DNA Repair and Mutagenesis*, Blackwell.

> This is a much more technical but very comprehensive account of the repair mechanisms associated with mutagenesis.

Griffiths, A. J. G., Miller, J. H., Suzuki, D. T., Lewontin, R. C. and Gelbart, W. M. (1993) *An Introduction to Genetic Analysis*, 5th edn, W. H. Freeman and Company.

> This gives a very comprehensive account of the mechanics of mutation in lower animals and in humans.

Neel *et al.* (1988) Search for mutations altering protein charge and/or function in children of atomic bomb survivors: final report, *Am. J. Hum. Genet.*, **42**, 663–676.

Weatherall, D. (1993) *The New Genetics and Clinical Practice*, Oxford University Press.

> Describes the results of mutation studies in humans.

Chapter 5

Skelton, P. (ed.) (1993) *Evolution: a Biological and Palaeontological Approach*, Open University/Addison Wesley.

> A remarkably comprehensive account of the topics covered here and many others.

Jones, S., Martin, R. and Pilbeam, D. (eds) (1994) *The Cambridge Encyclopedia of Human Evolution*, Cambridge University Press.

> A discussion of the evolution of our own species in more depth.

Chapter 6

Cann *et al.* (1987) Mitochondrial DNA and human evolution, *Nature*, **325**, 31–36.

Darley-Usmar, V. and Schapira, A. H. V. (1994) *Mitochondria: DNA, Proteins and Disease*, Portland Press.

> A more technical account of cytoplasmic inheritance in humans.

Griffiths, A. J. G., Miller, J. H., Suzuki, D. T., Lewontin, R. C. and Gelbart, W. M. (1993) *An Introduction to Genetic Analysis*, 5th edn, W. H. Freeman and Company.

> The story of how cytoplasmic genes were discovered.

Grivel (1989) Mitochondrial DNA: small, beautiful and essential, *Nature*, **341**, 569–571.

McConkey, E. H. (1993) *Human Genetics, the Molecular Revolution*, Jones and Bartlett.

> A good account of human disease and cytoplasmic inheritance.

Wallace, D. C. (1989) Mitochondrial DNA mutations and neuromuscular disease, *Trends in Genetics*, **5**, 9–13.

Chapter 7

Blackburn, E. (1991) Structure and function of telomeres, *Nature*, **350**, 569–572.

Thompson, M. W., McInnes, R. R. and Willard, H. F. (1991) *Genetics in Medicine*, 5th edn, W. B. Saunders and Co.

> Chapter 14: genetics of the immune system.
>
> Chapter 16: genetics of cancer.

Willet, W. C. (1994) Diet and health: what should we eat? *Science*, **264**, 532–537.

Chapter 8

Kerem, B., Rommens, J. M., Buchanan, J. A., Markiewicz, D., Cox, T. K. Chakravarti, A., Buchwald, M., Tsui, L. -C. (1989) Identification of the cystic fibrosis gene: genetic analysis, *Science*, **245**, 1073–80.

Riordan, J. R., Rommens, J. M., Kerem, B., Alon, N., Rozmahel, R., Grzelczak, J., Zielenski, J., Lok, S., Plavsic, N., Chou, J. L., Drumm, M. L., Iannuzzi, M. C., Collins, F. S., Tsui, L. -C. (1989) Identification of the cystic fibrosis gene: cloning and characterization of complementary DNA, *Science*, **245**, 1066–73.

Rommens, J. M., Iannuzzi, M. C., Kerem, B. -S., Drumm, M. L., Melmer, G., Dean, M., Rozmahel, J. L., Cole, J. L., Kenedy, D., Hidaka, N., Zsiga, M., Buchwald, M., Riordan, J. R., Tsui, L. -C., Collins, F. S. (1989) Identification of the cystic fibrosis gene: Chromosome walking and jumping, *Science*, **245**, 1059–65.

Tsui, L. -C. (1992) *Trends in Genetics*, **8**, 392–398.

Acknowledgements

We are grateful to the readers of this book, Hilary MacQueen and Jennie Simmons, whose comments on early drafts were important in preparing the final text, and to John Greenwood, Librarian, and Jean Macqueen, Indexer.

Grateful acknowledgement is made to the following sources for permission to reproduce material in this book:

Figures and Table

Figures 1.2, 1.6a, b, 4.4–4.6, 4.10, 4.19, Table 4.2, Figures 5.3, 6.1, 6.2 Griffiths, A.J.F. *et al.* (1993) *Introduction to Genetic Analysis*, 5th edn © 1993 W.H. Freeman & Co.; *Figure 1.7* Carter, C.O. (1962) *Human Heredity*, Penguin © 1962 C.O. Carter; *Figures 1.8, 2.33* Mange, A.P. and Mange, E. J. (1990) *Genetics: Human Aspects* © 1990 Sinauer Associates; *Figure 1.9* Dickerson, R.E. and Geis, I. (1969) *The Structure and Action of Proteins*, Benjamin/Cummings Publishers © 1969 R.E. Dickerson and I. Geis; *Figure 1.10* Hurry, S. (1965) *The Microstructure of Cells*, John Murray; *Figures 1.11, 1.14, 1.19* Scriver, C.R. *et al.*, *The Metabolic Basis of Inherited Disease*, McGraw-Hill Information Services, Marston Books; *Figure 1.12* Karlsson, S. and Nienhuis, A.W. (1985) *Annual Review of Biochemistry*, **54**, Annual Reviews Inc.; *Figure 1.13* Weatherall, D.J. and Clegg, J.B. (1981) *The Thalassaemia Syndrome*, Annual Reviews Inc.; *Figure 1.16* Raff, R.A. and Kaufman, T.C. (1983) *Embryos, Genes and Evolution*, Macmillan; *Figure 2.1* Stryer, L. (1988) *Biochemistry*, 3rd edn © 1988 W.H. Freeman & Co., after Kornberg, A. (1980) *DNA Replication*; *Figure 2.2* Mange, E.J. and Mange, A.P. (1994) 'Human chromosomes', from *Basic Human Genetics*, © 1994 Sinauer Associates, image courtesy U.K. Laemmli; *Figures 2.3a, b, 2.35* © M. Daker; *Figures 2.6, 2.16* McConkey, E.H. (1993) 'Genetic map of human chromosome 21', in *Human Genetics: The Molecular Revolution*, Boston, Jones and Bartlett; *Figure 2.7* Roberts, J.A. and Pewby, M.E. (1985) *An Introduction to Medical Genetics*, OUP; *Figures 2.9, 2.10b, 2.12* © Mari-Wyn Barley; *Figure 2.14* Watson, J.D., Gilman, M., Witkowski, J. and Zoller, M. (1992) *Recombinant DNA* © 1992 W. H. Freeman & Co.; *Figure 2.17* Ballabio (1994) 'Contiguous deletion systems', in *Current Opinion in Genetics and Development*, **7**(1); *Figure 2.25* Sall, T., Nilsson, N.-O. and Bengtsson, B.O. (1993) 'When everyone's map is different', *Current Biology*, **3**(9); *Figure 2.26* Moses, M.J., Counce, S.J. and Paulson, D.F. (1975) 'Synaptonemal complex complement of man in spreads of spermatocytes', *Science*, **187**, 31 Jan. © 1975 AAAS; *Figure 2.27* Maclean, N. (1989) 'Analyzing the genome', in *Genes and Gene Regulation* © 1989 N. Maclean; *Figure 2.28* Willard, H.F. (1990) 'Reviews: centromeres of mammalian chromosome', *Elsevier Trends Journal*, **6**(12), Dec. © 1990 Elsevier; *Figure 2.30* Dimmock, N.J. and Primrose, S.B. (1987) *Introduction to Modern Virology*, Blackwells; *Figures 2.32, 3.4, 3.5* Lewin, B. (1994) *Genes V*, OUP and Cell Press; *Figure 2.34* Page, D.C. *et al.* (1987) 'A restriction fragment detects highly conserved sequences ...', *Cell*, **51**(6), 24 Dec. *Figures 3.2, 3.11, 3.19, 3.24* Latchman, D. (1990) *Gene Regulation*, Unwin Hyman; *Figure 3.9* Lewin, B. (1990) 'Controlling prokaryotic gene expression', *Genes II*, OUP and Cell Press; *Figures 3.10, 3.25a, b–3.27* Singer and Berg (1991) *Genes and Genomes*, University Science Books; *Figure 3.12* Buratowski, S. *et al.* (1989) 'Five intermediate complexes ...', *Cell*, **56**(4), 24 Feb.; *Figure 3.14* Alberts, B. *et al.* (1989) *Molecular Biology of the Cell*,

Index